AF378916

Physical Red Team Operations

Physical Penetration Testing with the REDTEAMOPSEC™ Methodology

Written by
Jeremiah Talamantes

Edited by
Derek Sandbeck

PHYSICAL RED TEAM OPERATIONS

PHYSICAL PENETRATION TESTING WITH THE REDTEAMOPSEC™ METHDOLOGY

JEREMIAH TALAMANTES
CISSP, CCISO, CEH, CCENT, CHFI

Copyright © 2019 by Jeremiah Talamantes

Edited by Derek Sandbeck

All rights reserved. This book or any portion thereof may not be reproduced or used in any manner whatsoever without the express written permission of the publisher except for the use of brief quotations in a book review or scholarly journal.

First Printing: 2019

ISBN-13: 978-0-578-53840-2

This publication is designed to provide accurate and authoritative information in regard to the subject matter covered. The author has made every effort in the preparation of this book to ensure the accuracy of the information. However, information in this book is sold without warranty either expressed or implied. The author or publisher will not be liable for any damages caused, or alleged to be caused, either directly or indirectly by this book.

Library of Congress Control Number: 2019945783
Hexcode Publishing, Excelsior, MN.

Full-Force Red Team® is a registered trademark of RedTeam Security Training, LLC. REDTEAMOPSEC™ is a trademark of RedTeam Security Training, LLC.

Ordering Information:

Special discounts are available on quantity purchases by corporations, associations, educators, and others. For details, contact neobotnet@gmail.com

PHYSICAL RED TEAM OPERATIONS

DEDICATION

This book is dedicated to my beautiful family. Katie, I'm not sure how I got so lucky. You humble me with your kindness, thoughtfulness, and patience. You help me become a better person each and every day. You've got your work cut out for you!

To my sweet Max and Emmy, you have shown me that the greatest privilege in this world is to be your daddy. I hold you both in the warmest place in my heart, forever. I love you more than words can express.

Sha sha.

MORE FROM THIS AUTHOR

The Social Engineer's Playbook: A Practical Guide to Pretexting

Buy online on Amazon.com at https://amzn.to/2NENCI3

The Social Engineer's Playbook is a practical guide to pretexting and a collection of social engineering pretexts for Hackers, Social Engineers, and Security Analysts. Build effective social engineering plans using the techniques, tools, and expert guidance in this book.

Learn valuable elicitation techniques, such as Bracketing, Artificial Ignorance, Flattery, Sounding Board, and others. For more information about this book, please visit Amazon.com.

Buy now on Amazon.com:

https://amzn.to/2NENCI3

SPECIAL GIFT

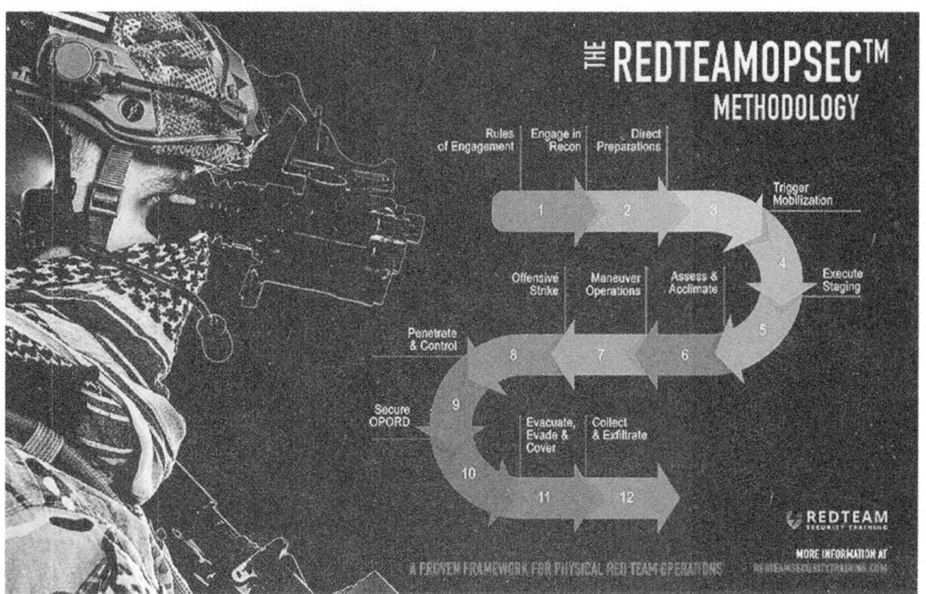

Get this FREE 11x17 REDTEAMOPSEC Poster

First of all, thank you for purchasing my book. As a small thank you, please visit the link below to get your hands on this snazzy REDTEAMOPSEC™ Methodology poster. Enjoy!

Visit this link:

https://redteamsecuritytraining.com/physical-red-team-book-gift/

TABLE OF CONTENTS

xvii

ACKNOWLEDGEMENTS

I want to thank my loving parents, Ray and Alma, and my brother Johnnie. I would also like to thank my father-in-law, Steveo, and my mother-in-law, Beth, for all of their love and support. Lastly, I would like to thank RedTeam Security and RedTeam Security Training for affording me the opportunity to pursue this endeavor.

https://www.redteamsecuritytraining.com

https://www.redteamsecure.com

PREFACE

This book is about physical red team operations, also referred to as physical penetration testing. Physical red teaming is not a new concept. However, it is practiced far too infrequently when compared to other forms of red teaming, even in today's times. Additionally, there aren't many books on the subject. In fact, I struggle to find more than a few even as I write. My goal in writing this book is to provide a guided approach toward physical penetration testing that wasn't available when I dove into it many years ago. I hope to bring about a grounded, repeatable and comprehensive structure using the REDTEAMOPSEC™ methodology – a wide-ranging 12-step framework for conducting physical red team operations.

My experience in red team operations started many years ago while working as an employee and as an independent security consultant. Throughout several facets of my career early on, I was fortunate enough to be exposed to red teaming before actually performing the work. Naturally, this helped me tremendously. Even though I had great successes, I felt there was something missing. This void traveled with me up until the point I founded RedTeam Security back in 2008.

While my consulting work had successful results, the lack of any known framework/methodology left me feeling somewhat lost. To operate with any level of consistency, I had to create my own framework. Hence, the REDTEAMOPSEC™ methodology, which is the core foundation of this book.

I hope this book provides red teams with the knowledge they need to further enrich their work and carry out physical red team operations with precision, confidence, and value.

TARGET AUDIENCE

This book was written for any person interested in learning more about physical security with a focus on physical red teaming operations. It is assumed the reader has a basic working knowledge about red team concepts and information security technology.

This book is recommended especially for security consultants, IT security analysts, blue teams, red teams, security managers, and CISOs. However, if you have an interest in securing your environment or testing other environments, this book is for you.

LIMITATION OF LIABILITY / DISCLAIMER OF WARRANTY

THE PUBLISHER AND THE AUTHOR MAKE NO REPRESENTATIONS OR WARRANTIES WITH RESPECT TO THE ACCURACY OF COMPLETENESS OF THE CONTENTS OF THIS WORK AND SPECIFICALLY DISCLAIM ALL WARRANTIES OF FITNESS FOR A PARTICULAR PURPOSE. NO WARRANTY MAY BE CREATED OR EXTENDED BY SALES OR PROMOTIONAL MATERIALS. THE ADVICE AND STRATEGIES CONTAINED HEREIN MAY NOT BE SUITABLE FOR EVERY SITUATION. THIS WORK IS SOLD WITH THE UNDERSTANDING THAT THE PUBLISHER IS NOT ENGAGED IN RENDERING LEGAL OR OTHER PROFESSIONAL SERVICES. NEITHER THE PUBLISHER NOR THE AUTHOR SHALL BE LIABLE FOR DAMAGES ARISING HEREFROM. THE FACT THAT AN ORGANIZATION OR WEB SITE IS REFERRED TO IN THIS WORK AS A CITATION, SOURCE OR OTHERWISE DOES NOT MEAN THAT THE AUTHOR OR PUBLISHER ENDORSES THE INFORMATION THE ORGANIZER MAY PROVIDE, OR RECOMMENDATIONS IT MAY MAKE.

BASIC TERMINOLOGY

TTPs (Tactics, Techniques, and Procedures) – An analysis of the patterns used by bad actors when performing attacks. "Tactics" is sometimes replaced with "tools" in the acronym.

Full-Force Red Teaming® – A patent-pending red team methodology created by the author, Jeremiah Talamantes, that involves simultaneous Physical Penetration Operations, Social Engineering, and Technology-based Penetration Testing.

Social Engineering – The act of manipulating people into performing actions or divulging information of your choosing through deceptive tactics. See also Elicitation.

Bad Actors – A generic title given to hackers, thieves, malicious employees, etc.

Operation – The term given to the entire physical red team project. Also referred to as a mission or an engagement.

Operators – A general title given to the skilled professionals who perform red team operations. They are also referred to as red teamers and team members.

Red Team Leader – The senior or lead position within a red team.

Stakeholders – These are the primary points of contact at the companies and organizations who hire for red team operations. They are also referred to as clients.

Attack Vectors – Any number of ways a target can become exploited, such as phishing, tailgating, phone phishing, baiting, etc.

Pretext – A falsified scenario and identity designed to deceive targets during an operation.

Pretexting – Merely refers to the act of carrying out a pretext.

Elicitation – Surreptitiously acquiring information from targets without them knowing.

Vulnerability – A known security weakness, such as a buffer overflow.

Exploitation – The act of taking advantage of a known security weakness or vulnerability.

Payload – The part of an exploit that executes malicious actions. For example, shellcode that elevates privileges in malware. This is not restricted to technical environments.

PII – Personally Identifiable Information, such as date of birth, social security number, etc.

Kit – One's tool bag consisting of software, props, hardware, and other physical tools.

Mission Success – The point when all red teamers have successfully met their mission objectives.

[CHAPTER 1]

INTRODUCTION TO PHYSICAL RED TEAMING

The computer hacker crouched low in thick brush on a cold December night, just beyond the fence line of his target – a massive U.S. oil refinery. Wearing night-vision goggles and dressed in black, he swung a rubber mallet into the dirt, trying to produce vibrations to distract the plant's ground-penetrating radar system. He swung again and again. Flashlights emerged from a distant building, then disappeared. Soon a train roared by, providing the cover his team needed. Quickly, two more men appeared from the shadows. They threw a wool blanket over a 16-foot barbed wire fence, climbed over and rushed to a small building housing the facility's vital computer controls. The door had an electronic lock, a badge reader, and a plate to thwart lock picking. But the intruders caught a break. The door didn't sit properly in its frame, leaving just enough space to shimmy it open.

Within moments, they had planted a small device, about the size of a credit card, designed to begin penetrating the refinery's controls systems. "Bingo!" crackled from the radio inside a white SUV adorned with a phony logo of the refining company, some 200 yards away. From there, Jeremiah Talamantes gave the signal to leave – "Rabbit!"

As the other hackers hopped in the van, the driver's nerves calmed. Then a stark reality set in.

"We've used a couple hundred dollars in gear, and we were able to break into a refinery without anyone knowing," said Talamantes, president and managing partner of RedTeam Security in Minnesota.

"The implication is pretty devastating."

Talamantes was hired by the refinery to test its defenses against cyberattacks and, like so many others, the mission was way too easy. Despite the refinery's remote location, fencing, high-tech sensors and security team, his team was able to infiltrate its network and potentially wreak havoc.

[...]

As Talamantes' refinery caper shows, hackers don't have to limit themselves to the internet to break into computer networks. With long-range cameras, they can spend days watching workers entering through front doors, so they can mimic their behavior and exploit weak spots to get inside, Talamantes said. Before Talamantes and his team raided the oil refinery in December, they staked out the company's corporate offices. They watched employees at nearby coffee shops and restaurants, managing to steal and clone badges.

Talamantes said he tries to stay within the bounds of what real hackers can do with a modest investment. In the refinery raid, his team carried only a small amount of gear, including a laptop, lock-pick set, and a $35 device to tap the computer systems, all available on Amazon.

They used two 16-foot ladders, which they returned to Home Depot for a full refund, a set of four two-way radios, and lock picks. Over the course of his career, Talamantes said, such tests have found plenty of security weaknesses, cyber and otherwise, that should worry the energy industry.

But the scariest part, he said, is that so much of hacking is low-tech, requiring little expertise.

"Anyone can do these types of things."

Eaton, Collin. "Put to the Test, Cybersecurity Experts Easily Infiltrate Energy Companies' Networks." Houston Chronicle, 6 Feb. 2018, www.houstonchronicle.com/business/article/Put-to-the-test-cybersecurity-experts-easily-10989830.php?t

To many people, this may read like a scene straight out of some hacker-spy movie mashup. Perhaps it should've started with, "It was a dark and stormy night." No, not quite. In reality, this was an article written by Collin Eaton of the Houston Chronicle recounting a real operation by my team at RedTeam Security (https://www.redteamsecure.com). So for now, let's all breathe a big sigh of relief that this was just a test and nobody was hurt.

I apologize for the dramatic intro, but I believe it illustrates an important point. What this article offers is a small glimpse into the world of Physical Red Teaming and to what extent companies need to take steps to protect themselves from bad actors. Today, bad actors assume many forms, and their malicious plans range far greater than they did even five years ago.

Are you wondering about the security posture of your company or organization right about now? Does this sound like something you should consider for your own organization? Or does it sound like an interesting career path? Well I hope your inner wheels are turning, if only just a tiny bit.

In this book, my hope is to provide readers with an understanding of physical threats and how they seep into other threat domains, cyber and social alike, and have an overall impact on a company. More importantly, the focus of this book is to provide a comprehensive guide to executing physical red teaming with accuracy and effectiveness.

As a side note, I will use the terms "physical red teaming" and "physical penetration testing" interchangeably.

Overview

Physical red teaming is an exciting way to earn a living, and the career field itself is emerging. To date, there isn't a college course that will get you started in physical penetration testing straight out of school. Methodologies are piecemeal, tactics, techniques and procedures (TTPs) are vague, execution is often sloppy, objectives are misaligned, and results are inconsistent. Boutique training, like the kind provided by RedTeam Security Training (www.redteamsecuritytraining.com) and books like this one are your best resource. The focus of this chapter is to establish a basic foundation on physical red teaming, whether you have some experience or none.

In short, a physical penetration test is an authorized simulated physical attack on an organization's physical security in an effort to evaluate their security posture and, ultimately, help improve it. To identify security weaknesses, measure security posture, and offer ways to improve it is a shared goal of just about any penetration test. Governments, companies, and organizations make use of penetration testing in order to improve the effectiveness of their security controls. There are, however, some stark differences between physical penetration testing and the other variants.

What makes physical red teaming unique is the absence of (most) computing technology as a target. When you hear someone mention penetration testing, most people immediately think technology is the intended target, such as computers, web applications, firewalls, networked devices, and so on. That certainly doesn't mean that computing technology is never used or targeted during physical penetration tests. Instead, we use this simple distinction to convey the things we primarily aim to test. These often include motion detectors, security cameras, employees, access control scanners, security fences, locks, security personnel, and other controls and technologies intended to keep physical assets secure. So to clarify, computer systems and related technologies are utilized and exploited in an effort to further physical security operations.

Physical security operations require an appreciation for complex puzzles and out-of-the-box problem-solving. Fortunately for many traditional penetration testers and hackers, this is an easy transition. It's becoming familiar with physical security methodologies, threats, vulnerabilities, and tools that become the crux. I can speak from experience; this was one of the most trying aspects for me.

In this book, I hope to build upon your hacker mindset and bridge the gap of know-how by showing how to execute physical penetration tests professionally.

History

Before we dive into the history of physical penetration testing, let's first take a look at its origins. What we know as penetration testing, whether it be technical or physical, began as an offshoot of Red Teaming. Red Teaming has many other names, such as Red Cell, Opposing Force (OPFOR), Tiger Team, and so on. It was developed by military organizations, particularly the U.S. military, as a way to challenge an organization's defenses, discover new vulnerabilities, and to improve and measure its effectiveness by assuming an adversarial role. An independent team assumes the role of the adversary and is referred to as the red team, also known as "the attackers" or "bad guys." Consequently, the team charged with defending against attacks from the Red Team is called the Blue Team.

In the military and intelligence world, early red teams were generally focused on utilizing a red team methodology as an alternative approach to problem-solving. In the ancient Indian military somewhere around 320 AD, battle strategies were played out on what might have looked like modern board games. These tools provided military commanders with a way to play out enemy moves and strategize. Tabletop exercises in today's military are the result of many evolutions of the red team approach.

In the early 1960s, Robert H. Davis published an article titled, "Arms Control Simulation: The Search for an Acceptable Method."

While the article's content is irrelevant to the subject of this book, it is one of the first publicly documented examples of red team strategy in play. In the late 1960s, a few people speaking at a conference from the RAND Corporation and the NSA coined the term "penetration" to describe an attack against a computer system. Soon the term penetration would evolve into "penetration testing."

As time wore on, penetration testing would grow to be divided into categories like network Penetration Testing, Application Penetration Testing, and, our favorite, Physical Penetration Testing.

Types of Penetration Testing

I felt it worthwhile to spend just a few cycles writing about the different types of penetration testing. As we've just learned, penetration testing has grown to be divided to include other focus areas outside of just physical penetration testing. Let's take a minute to explore the common ones as they pertain to information security.

Network Penetration Testing

One of the first types to enter the scene was network penetration testing. Although it wasn't the very first, it was the variant that made the biggest impact early on. As it states, the network penetration test is an authorized simulated attack on computer systems in an effort to evaluate their cumulative security posture. Computer systems typically are comprised of servers, network devices (switches, hubs), appliances (firewalls, VPNs), and just about any device on the external perimeter that holds an IP address.

As network penetration testing and security threats evolved, security administrators began to look to internal networks and the systems that reside there. As a result, further division occurred. And soon they began to refer to them as external network penetration tests and internal network penetration tests, respectfully.

At its core, network penetration testers hunt for system and service misconfigurations, extraneous services, service-specific vulnerabilities, product-specific vulnerabilities, access control issues, and some operating system vulnerabilities, to name a few. The delineation between network penetration testing and our next type, application penetration testing, can best be separated using the 7 layers of the OSI model. Network penetration testing typically includes the entire media layer (layers 1 to 3) and sometimes layers 4 and 5 of the host layers. Application penetration testing includes layers 7, 6, and sometimes lower layers.

For more information about network penetration testing, visit the Penetration Testing Execution Standard (PTES) website at www.pentest-standard.org.

Web Application Penetration Testing

Web application penetration testing followed network penetration testing. A web application penetration test concentrates only on evaluating the security of a web application. The process involves an active analysis of the application for any weaknesses, technical flaws, or vulnerabilities. There are many types of web app vulnerabilities, such as SQL injection, privilege escalation, session fixation, reflected cross-site scripting, weak cryptography, and many more.

For more information concerning web application security, please visit the Open Web Application Security Project (OWASP) website at www.owasp.org.

These days, web application penetration tests are just as common as those that home in on system and network environments. This was not the case during the early 2000s. On the topic of emerging focus areas, mobile applications that natively run on smartphones are a growing sector.

Mobile Application Penetration Testing

Applications developed for mobile smartphones have exploded in the past several years. Games, business tools, self-help, utilities-- you name it, and there is probably a mobile app for it. With all these apps being developed on ever-evolving smartphone platforms at breakneck speeds, ensuring a securely built app is extremely problematic. As a result, mobile applications today represent a great risk for end-users. Mobile application penetration testing is in high demand for obvious reasons.

The overall goal for mobile penetration testing is no different that the aforementioned penetration testing types. However, the threats and vulnerabilities differ slightly. Security issues tend to lie heavily in the areas of application local data storage, authentication/authorization, cryptography, mobile platform interaction (Android/iOS), and more. Thankfully, there are many resources available on the study of mobile application security.

As always, OWASP is a great resource for application-related security. Please see their website for lots of great information: https://www.owasp.org/index.php/OWASP_Mobile_Security_Testing_Guide.

Wireless Penetration Testing

Wireless network penetration testing is sometimes rolled into a network penetration test because of its relation to networks and networked systems. However, I'll briefly cover it here.

Most wireless pen tests target susceptibility to sniffing packets by decrypting poorly developed Wi-Fi encryption algorithms (WEP), man-in-the-middle attacks, rogue AP testing, client attacks, DoS attacks, etc. There are some great tools available, such as the Wi-Fi Pineapple, that make testing easier and even fun.

Wireless penetration testing isn't limited to 802.11x. Other platforms, like Bluetooth and RFID, are commonly in-scope as well. A well-rounded wireless penetration test should cover all platforms in use by the client.

For more information about wireless penetration testing, visit the Penetration Testing Execution Standard (PTES) website at www.pentest-standard.org.

IoT Penetration Testing

With the advancement of the Internet of Things (IoT), it was only a matter of time before security professionals caught up. IoT devices range from embedded devices in cars to smart cities to watches to thermostats. Ironically, IoT hasn't introduced any drastically new technology, but it meshes different technologies in new and complex ways.

What's new and novel about IoT are the mashups of different communication protocols, architectures, languages, and operating systems. Most traditional penetration testers have been confronted with Windows and Linux x86/x64-bit systems over the familiar TCP/IP protocol. Once IoT is added in the mix, there will be many new protocols (ZigBee, NFC, SDR) and architectures for which there is little security knowledge and few penetration testing tools.

For more information concerning IoT security, please visit the IoT Security Foundation's website at www.iotsecurityfoundation.org.

How It's Used Today

As previously mentioned, physical penetration testing began as a derivative of red teaming within the military. It eventually expanded into the private sector, initially by private contractors with close ties to the U.S. Government. In contrast, today, there are many organizations in various industries including, but not limited to, manufacturing, major retail, critical infrastructure, wealth management, high net worth individuals/families, banking, healthcare, data centers, and more.

Unless you are in the military, law enforcement, or are a private contractor, the people at the facilities you will be tasked to infiltrate will likely not be armed, and there will be little risk of severe danger. Of course, I say this with a bit of restraint because my team has tested very secure facilities where the risk of severe danger was high. Fear not, testing under these circumstances is rare.

What physical red teamers aim to do is exploit the physical security controls intended to protect assets that are important to the client via simulated attacks. It's important to note that what is considered important to the client will probably not be immediately evident on day one of planning. Arriving at that point will take time and require testers to tease that information out of a client. We will cover that in the planning portion of this book.

Successful physical red teamers use comprehensive, repeatable approaches to their operations to ensure consistency, value, and comprehensiveness. The process used in this book includes a multi-step process for accurate scoping, resource planning, execution, and reporting. Before any physical operation can begin, these critical preparatory stages must be reached, understood, and approved by all involved.

In order to deliver on the promise to produce a step-by-step guide to physical red teaming, it's important to systematically divide a battle-tested execution methodology for easier consumption. As a result, the majority of this book's content will encompass a 12-step process designed to take you from the Rules of Engagement to exfiltration.

REDTEAMOPSEC Methodology™

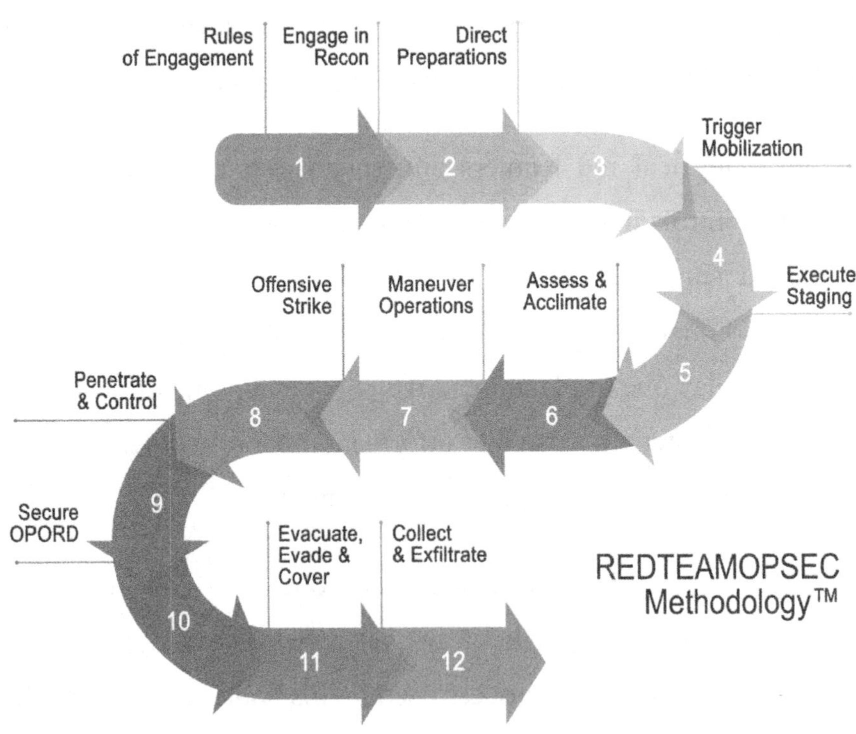

Figure 1. REDTEAMOPSEC Methodology™

The REDTEAMOPSEC methodology provides a quick glance at my approach for conducting physical red team operations and will serve as the chapter outline for the remainder of this book. I developed this methodology in response to a lack of any real and relevant framework on the subject of physical red teaming. As you might have already noticed, REDTEAMOPSEC is an acronym for each of the twelve crucial and distinct steps in the physical red teaming process.

Here are the 12 steps in the REDTEAMOPSEC methodology:

1. **R**ules of Engagement

2. **E**ngage in Reconnaissance

3. **D**irect Preparations

4. **T**rigger Mobilization

5. **E**xecute Staging

6. **A**ssess & Acclimate

7. **M**aneuver Operations

8. **O**ffensive Strike

9. **P**enetration & Control

10. <u>S</u>ecure OPORD

11. <u>E</u>vacuate, Evade, & Cover

12. <u>C</u>ollect & Exfiltrate

REDTEAMOPSEC was founded upon a very strong military influence leveraging the countless benefits of military strategy. In this book, there are also many images depicting firearms and soldiers. These images were chosen for their applicability toward the subject matter, and no disrespect is meant to our men and women in the armed forces.

That said, the REDTEAMOPSEC methodology won't make much sense right now, and that's okay. A chapter will be devoted detailing each of the following twelve steps in chronological order. My hope is that you'll be able to use this book as somewhat of a field manual enabling you to quickly page to your chapter of interest. As a result, some chapters will be longer than others while others will be shorter. However, in the long run, I think this will be a benefit more than a publishing faux pas.

Without further ado, let's dive right into the REDTEAMOPSEC methodology.

[CHAPTER 2]

RULES OF ENGAGEMENT

A physical red team operation should not begin unless the testing team has a clear understanding of the target's threat profile. Identifying the target's exposure factors ultimately enables operators to develop an operation leveraging realistic TTPs and level of sophistication against threats the target will likely face. There is more on this in Chapter 14, "Full-Force Red Teaming."

To pause for a moment and state the obvious, threat profiling is not the very first step in an operation. Depending upon if the red team is an internal team or if the team is part of a security firm will ultimately play an important role in what the very first steps in launching an operation truly are. It's not the goal of this book to focus on the front end of the engagement process. But having an agreed-upon scope, objectives, and a prepared plan make up those crucially important pre-engagement requirements.

Overview

Before a physical red team operation can begin, there must be an understanding between the testing team and the client on some very important things; namely, those things that revolve around certain specifics pertaining to TTPs and how they will be carried out. For this, we start at the first phase in the REDTEAMOPSEC methodology called, Rules of Engagement (RoE). The RoE is a document that outlines the entire operation from start to finish at a high level. It communicates important information and milestones to the client and much more.

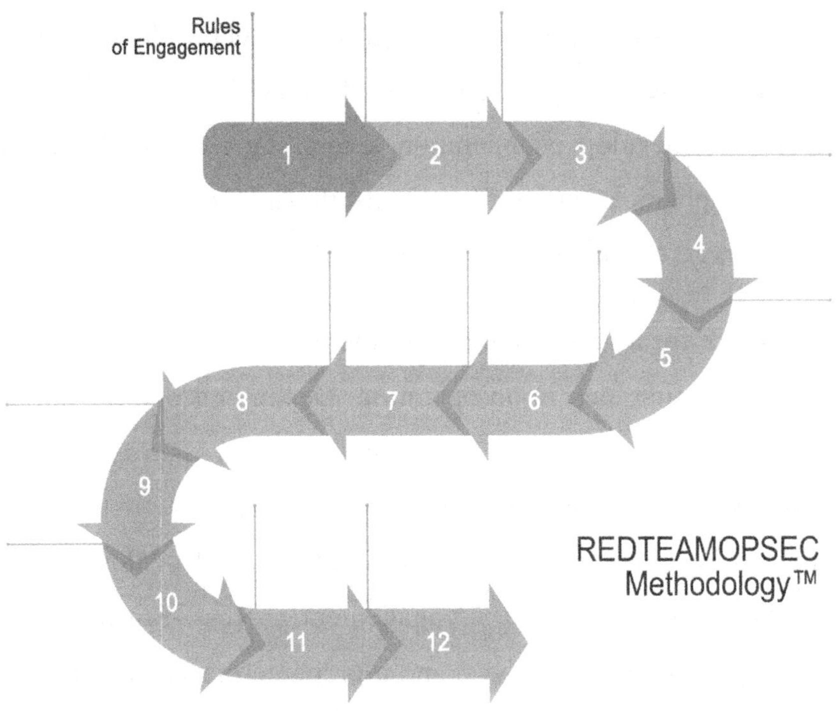

Figure 2. Rules of Engagement Phase

The Rules of Engagement has a few primary functions. Among those functions are to signify which TTPs and targets, from a high level, are fair game. Equally as important, it also signifies which TTPs and targets are prohibited, or out of scope Some TTPs may indirectly or directly cause damage to property. Damage to client property is sometimes considered out of scope and is often stated as such in the RoE. For example, "Any and all damage to client company property (e.g.: locks, doors, windows) is strictly prohibited."

In contrast, *some* damage to client property may be permitted by the client because it may be a realistic threat an attacker would use against them. This is precisely where the value of an RoE is realized.

A client may be okay with damage to certain locks and windows but not to others. Therefore, it is important to have this understanding clarified and agreed upon in writing. The key here is to be as verbose as possible to avoid confusion.

Damage to Property

To step back a moment, I'm not encouraging all-out physical damage to property on each and every operation for the sake of breaking things. The concept of breaking windows during a security test, for example, may seem unconventional to most clients. But what we should remind them of is the fact that bad guys don't play by the rules and we, as red teamers, should push the envelope to mimic their behavior as closely as possible.

We as physical red team operators must propose unconventional test scenarios to clients where it makes sense, and, of course, there has to be some solid reasoning behind the decision to damage property. Finding those situations is not always obvious, so I've put together some guidance to help that process along.

Please see the numbered list below when considering TTPs that involve potential or certain damage to client property.

1. The likelihood that a bad actor would use the same TTPs is significant enough to warrant potential or certain damage.

2. The negative impact a bad actor could incur using the same TTPs is significant enough to warrant potential or certain damage.

3. The use of damaging TTPs will add value and 'realism' to the operation.

4. TTPs utilize the same level of sophistication as a likely bad actor.

5. TTPs utilized are commensurate with the asset's value.

6. The client is aware of the asset's monetary value, business value, and downstream business function should the asset become broken or unavailable.

7. Both you and the client are in full agreement with chosen TTPs, assets, and how the asset will be fixed or reimbursed.

The goal of item #1 is intended to operate as a quick sanity check to ensure the threat is significant enough to test with the expectation of damage and the selected TTP is something a bad actor would actually utilize. We want to avoid overcomplicating TTPs, especially if the bad actor might choose a simpler route.

Item #2 is also a great thought exercise. Here we want to be sure the estimated impact is worthy of the resources it requires to test. For instance, most convenience stores *expect* kids to steal packs of bubble gum. It is considered a cost of doing business. But unless kids are stealing them by the pallet, the threat isn't impactful enough to install costly security controls or hire red teamers. Test only where the impact, and likelihood, for that matter, are significant enough to warrant the effort.

A TTP that is likely to cause damage shouldn't be carried out unless the client sees it as valuable to the operation. Breaking a physical control that is inexpensive or easily replaced should not be the sole factor in the decision-making process. Testing newly identified vulnerabilities or retesting old vulnerabilities with updated TTPs is a great way to add value. Doing this gives additional perspective and likely more food for thought.

Item #4 is one of my biggest pet peeves. Over eager red teamers like to over-engineer TTPs to be more like the fictional spy character, Ethan Hunt, than anything. I call this the Mission: Impossible Effect. The term **tacticool** also comes to mind. Simply put, physical security controls should be tested using the same kit and TTPs that a likely bad actor would use. TTPs and kit must be commensurate with the level of sophistication of the bad actor and the physical security control at hand. Any over exaggeration serves only to damage a team's reputation and the value of the operation.

Before making a recommendation to your client to use a damaging TTP, be sure your client has a good handle on the asset's value and any potential downstream impact. Naturally, you won't have these answers, but it is crucially important to walk the client through the brainstorming process. Something as simple as breaking a window may have complicated downstream effects. I have found it best to raise three important points during these discussions.

- Monetary cost (How much does it cost? Can it be replaced? How long to replace it?)

- Business functionality cost (What are its direct business implications? Loss of availability? Impact to Integrity? Impact to Confidentiality?)

- Downstream impact (Brainstorm potentially unforeseen downstream impact)

If it makes sense to use the TTP and the client gives the green light, the next step is to document this in the RoE. Sometimes an operation changes mid-course and an opportunity to use such a TTP becomes valuable. If so, it's critical to document the details in written format for clarity and non-repudiation purposes.

Below is an editable Microsoft Word template of the Rules of Engagement.

For an editable Rules of Engagement (RoE) template in Microsoft Word, please visit the following URL:

https://www.redteamsecuritytraining.com/physical-red-teaming-book

I will provide a brief outline of the same RoE below. However, I strongly encourage readers to visit the link instead, as the document will be updated and improved upon periodically.

RoE Outline

The numbered list to follow is an example of a high-level outline of a Rules of Engagement deliverable. I recommend using the existing components as a bare minimum and to modify as necessary. Again, this is a client-facing document that is presented and discussed at the start of every operation. Since this is one of the very first operational documents produced, it is probably going to be updated often as the operation progresses. Because of this, the client stakeholders must have complete visibility and be required to review and approve any changes as they happen.

Here are the major components of the RoE:

1. Client Name

2. Client Contacts

3. Project Contacts

4. Red Team Members & Roles

5. Target Location Address(es)

6. GPS Coordinates (Each location)

7. Operation Objective(s)

8. Target Control(s)

9. Out-of-Scope Control(s)

10. Out-of-scope TTP(s)

11. Damage Causing TTP(s)

12. Additional Notes

13. Document Change History Table

    a. Change Description & Date

    b. Client Change Review/Approval Signature & Date

Once again, I recommend readers visit the link published earlier for more information and a modifiable RoE template.

The RoE is a living document and must be updated as often as necessary and shared with client stakeholders and red teamers. Making use of an online client-facing portal to allow for easy document sharing is highly suggested. My team and I use such a portal that will sends alerts to all each time the RoE is updated. This gives clients an opportunity to review, ask questions, and provide written approval.

The key to success during any engagement is having a well-documented RoE that keeps clients updated and team members focused. I can't stress this point enough. It will certainly be an exercise in communication and follow through, but it is a necessity that clients will grow to love.

[CHAPTER 3]

ENGAGE IN RECONNAISSANCE

In this chapter, I will provide a quick overview of the concept of reconnaissance (recon) and how it should be carried out during physical red team operations. I will propose a couple of high-level, tactical approaches to take toward reconnaissance that I believe make the process more systematic, effective, and repeatable. Finally, I will close out the chapter by providing a list of must-have tactical surveillance gear my team and I use every day.

Overview

Reconnaissance is a mission to obtain information by visual observation or other detection methods, about the activities and resources of an enemy or potential enemy, or about the meteorological, hydrographic, or geographic characteristics of a particular area (Reconnaissance (US Army FM 7-92; Chap. 4).

A successful red team operation would not be possible without a solid foundation of actionable intel about the target or targets. What kind of intelligence? The location of security cameras, entrances, checkpoints, guard huts, and motion sensors are just a small example. Engaging in planned reconnaissance missions aimed at discovering these items is what this chapter is all about.

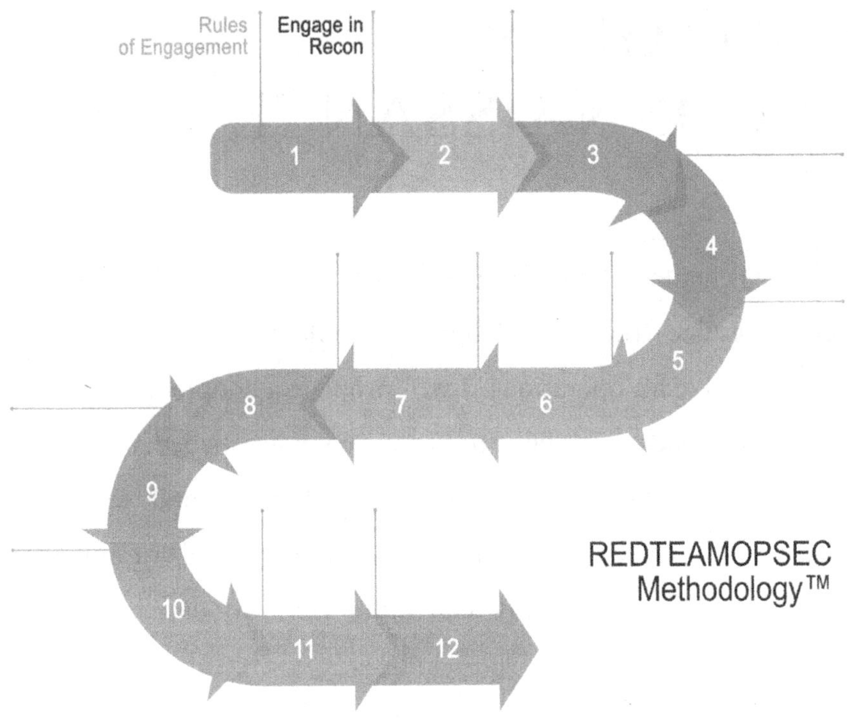

Figure 3. Engage in Reconnaissance Phase

As you've probably already realized, reconnaissance is a military tactic heavily used during any number of military operations. It is extremely useful in exploring areas across enemy lines in an attempt to gain useful information about an enemy's position, combat strength, terrain, weapons, etc.

Obviously, we are not engaging in military warfare here, but incorporating military TTPs during our physical red team operations has been paramount to the success of my team's operations. Therefore, this book will be heavy on military-themed concepts for their added benefits.

In this chapter, we will make use of an adapted version of military-themed reconnaissance tactics to help us obtain information about our targets in the same way bad actors might.

Before Getting Started

A few things must be in place prior to the start of a recon engagement. Some of these practices, such as inter-team communication during recon, might be altogether new to readers and may require some level of introduction. As a result, the subsections to follow aim to shed light on these critical components.

Red Team Leader

The hierarchy for small red teams is usually very flat. However, every team should be made up of two or more red team operators and at least one red team leader. Here are some of the basic responsibilities of a red team leader:

- Knows the mission at-hand completely and thoroughly

- Serves as primary communicator with client

- Serves as primary communicator between team operators

- Certifies team readiness

- Responsible for the actions of the operators

- Commands the operators in the field

- Determines mission success

Generally, the red team leader is the most experienced on the team. She must possess excellent communication, organizational, and tactical skills.

Project Repository

Critical to any engagement are project documents, spreadsheets, project notes, evidence, and so on. By now, we already have the Rules of Engagement and a fairly good understanding of the operation as a whole. As we progress through the REDTEAMOPSEC methodology chain, additional project artifacts will be created, shared, and updated. Therefore, it is imperative to establish a centralized and secure repository for disseminating and communicating in writing.

I encourage using an online portal system expressly designated for document sharing coupled with advanced features to notify users and provide a means to comment and collaborate. If this isn't immediately available, one could make do by using Google Drive, Google Sheets, and Google Docs.

Communication

It goes without saying, but ineffective communication will ruin any and every engagement. So to start on the topic of communication, we will focus on two types:

- Client Communication

- Inter-team Communication

Client Communication

Expectations should be set in advance on what kind of information the client might expect to receive, approve, or collaborate on. In the beginning, this will likely be the RoE. However, clients should have a basic understanding of the many different types of documents that could be shared and what actions they should take in response, if any. For example, updates to formal documents or agreements, such as an RoE, will require review and approvals. Intel the red team uncovers on targets may only require a client's review. Photos the red team takes during recon missions may not require any action on behalf of the client. In any event, we don't want our client to become confused about what to do and how to respond to the many pieces of information they find in their possession.

A cadence of communication should be established and understood between client and red teamers. This becomes more important when a red team is actively engaged while deployed onsite during a recon mission, for example. This type of communication is most often conducted by phone, text, email and radio respectively. Therefore, the client should be informed and expect to receive and respond to a high volume of communication from the red team during reconnaissance missions.

There should also be a designated list of contacts the red team communicates with during such recon missions. This communication happens during all hours of the day. Thus, these designated contacts must be available by phone, at a minimum, in the event something important is discovered or if something goes sideways during the recon mission. Generally, if something goes awry during recon, it usually means the recon team was compromised. In other words, an employee, bystander, or third party may have seen the recon team doing something suspicious, preventing the team from continuing.

To aid in client communications, a section in the RoE is often designated to define who the client assigns as its contacts along with their contact information and role. An additional piece of documentation called an Authorization Letter (aka: Get Out of Jail Free Card) will further describe the contact/escalation list along with additional information. The Authorization Letter is something we will cover in greater detail later in this chapter.

Inter-team Communication

Information designated as client-facing should be communicated through the red team's document repository. But much of the inter-team communication can and should happen in a team meeting or series of team meetings. Any output from those meetings should be uploaded to a document repository for internal use. On that note, let it be known that not every piece of documentation needs to be reviewed or shared with the client. This usually amounts to internal strategizing sessions and team planning estimates. That information can be limited to internal use only.

Inter-team communication, from resource planning to strategizing, will be gathered throughout the REDTEAMOPSEC phases. That said, a great deal of that work often occurs in the early phases of recon planning and during the execution phase.

Whenever a team meeting or discussion occurs, I highly recommend taking notes. I have found myself in many situations where my team rehashes topics that were previously discussed. Taking notes and sharing them with the team will keep them informed and more focused.

Here are some key points to capture during inter-team meetings, strategizing sessions, and discussions:

- Strategy ideas

- TTP planning

- Time constraints and travel

- Resource planning considerations

- Reconnaissance vantage points

- Risk areas for bystander detection

- Staying in alignment with objectives

- Recon equipment needs

Equipment

Equipment requirements will change from one recon mission to another. Even during the same engagement. Unfortunately, there is no one-size-fits-all solution. But what I will offer here is a list of equipment that my team and I tend to use on nearly every recon mission.

Most recon missions boil down to these important steps: Contact, Conceal, and Capture, what I call the **Recon C^3 Method**. We will talk more about the Recon C^3 method later in this chapter.

For now, here is a list of essential equipment my team uses on nearly every recon mission:

<u>Contact:</u>

- MOLLE tactical vests to help carry essential gear on your body

https://kit.com/redteamtraining/red-team-physical-penetration-testing-equipment/solomone-cavalli-tac

- Radios for on the ground communication between red teamers

https://kit.com/redteamtraining/red-team-physical-penetration-testing-equipment/midland-gxt1000vp4-3

- In-ear headset to help keeps hands more mobile

https://kit.com/redteamtraining/red-team-physical-penetration-testing-equipment/hde-2-pin-covert-in

- Tactical ripstop pants (in various colors) to prevent minor injury from terrain

https://kit.com/redteamtraining/red-team-physical-penetration-testing-equipment/5-11-tactical-74003

- Tactical ripstop shirt (in various colors)

https://kit.com/redteamtraining/red-team-physical-penetration-testing-equipment/5-11-tactical-72002

- Tactical day bag for carrying equipment

https://kit.com/redteamtraining/red-team-physical-penetration-testing-equipment/huntvp-10l-mini-dayp

- Compass to aid in navigation for wide open spaces

https://kit.com/redteamtraining/red-team-physical-penetration-testing-equipment/x-plore-gear-emergen

- Timekeepers for the recon team

https://kit.com/redteamtraining/red-team-physical-penetration-testing-equipment/hah-men-s-led-multif

- A decent multitool

https://kit.com/redteamtraining/red-team-physical-penetration-testing-equipment/gerber-dime-multi-to

Conceal:

- Balaclava

https://kit.com/redteamtraining/red-team-physical-penetration-testing-equipment/self-pro-balaclava

- Durable boots for various types of terrain

https://kit.com/redteamtraining/red-team-physical-penetration-testing-equipment/5-11-tactical-a-t-a

- Durable gloves for navigating over barbed wire or other rough obstacles

https://kit.com/redteamtraining/red-team-physical-penetration-testing-equipment/reebow-tactical-mili

- A thick wool blanket to help get you over that barbed wire fence

https://kit.com/redteamtraining/red-team-physical-penetration-testing-equipment/military-outdoor-clo

- Mylar blankets for deceiving heat-sensing security controls

https://kit.com/redteamtraining/red-team-physical-penetration-testing-equipment/primacare-hb-10-emer

- Head lamp with red light to enable covert movement

https://kit.com/redteamtraining/red-team-physical-penetration-testing-equipment/luxolite-led-headlam

- Tactical and portable red torch

https://kit.com/redteamtraining/red-team-physical-penetration-testing-equipment/turnraise-1200lm-xm

- Wireless endoscope for peeking around corners undetected

https://kit.com/redteamtraining/red-team-physical-penetration-testing-equipment/depstech-wireless-en

Capture:

- A tough laptop (Mac or Windows) to enable enhanced recon capture, etc.

https://kit.com/redteamtraining/red-team-physical-penetration-testing-equipment/mobiledemand-lc-rug

- Camera with excellent optical zoom

https://kit.com/redteamtraining/red-team-physical-penetration-testing-equipment/nikon-coolpix-p900-d

- Tripod enabling tight zoom shots

https://kit.com/redteamtraining/red-team-physical-penetration-testing-equipment/sunpak-620-020-tripo

- Night vision binoculars for spotting movement and infrared cameras and detectors

https://kit.com/redteamtraining/red-team-physical-penetration-testing-equipment/solomark-night-visio

- Small body worm camera (GoPro)

https://kit.com/redteamtraining/red-team-physical-penetration-testing-equipment/gopro-hero7-hero-7-b

- Head worn camera mount to enable hands-free movement

https://kit.com/redteamtraining/red-team-physical-penetration-testing-equipment/amazonbasics-head-st

- Binoculars

https://kit.com/redteamtraining/red-team-physical-penetration-testing-equipment/bushnell-powerview-s

- Thermal imaging camera for iPhone

https://kit.com/redteamtraining/red-team-physical-penetration-testing-equipment/flir-one-thermal-ima

- Discreet eye-glasses camera

https://kit.com/redteamtraining/red-team-physical-penetration-testing-equipment/wiseup-16gb-1920x108

- Capturing signals (Wi-Fi) intelligence

https://kit.com/redteamtraining/red-team-physical-penetration-testing-equipment/alfa-wifi-antenna-18

- Discreet mini camera pen for capturing recon up close

https://kit.com/redteamtraining/red-team-physical-penetration-testing-equipment/magendara-mini-camer

- All-weather field notebook

https://kit.com/redteamtraining/red-team-physical-penetration-testing-equipment/rite-in-the-rain-all

- All-weather writing utensil

https://kit.com/redteamtraining/red-team-physical-penetration-testing-equipment/1069028-rite-in-the-rain-all

Before embarking on a recon mission, the team should use a Load Out List to list and keep track of necessary equipment. Check out the Load Out List template below.

For an editable Load Out List template in Microsoft Word, please visit the following URL:

https://www.redteamsecuritytraining.com/physical-red-teaming-book

Environmental situations, such as urban settings vs. rural settings and day vs. night, will ultimately determine the extent to which these tools are relevant. Keep in mind this is not a full and complete list, however this is good enough to prepare any recon team from the get-go.

Planning Recon Missions

Though the specifics of a reconnaissance mission always vary, performing them can and should be carried out in a systematic process. No set of unique recon goals or objectives should completely deviate from a solid methodology.

From a high level, the C.O.V.E.R.T. Recon Method™ (COVERT) is a system that enables the recon mission process to happen repeatedly with consistency and confidence.

COVERT Reconnaissance Method

As you can see in Figure 4, the COVERT recon methodology is a very straightforward six-step process. Its best use case might be to use it as a high-level operational guide to red teams.

Later in this chapter, we will cover another method called the Recon C^3 method, whose intent is to help tactically guide red teams during the execution phase of a recon mission. But for now, the COVERT recon methodology can be used as a valuable tool to distill the sometimes complex process of carrying out reconnaissance missions.

In previous years, I struggled to produce consistent results from operation to operation. Eventually, I discovered one of the primary reasons was due to a fickle process I was using to gather information. COVERT recon helps smooth those edges.

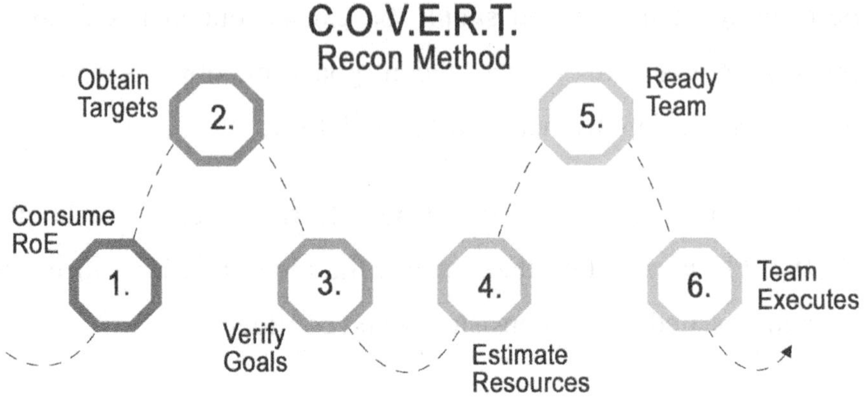

Figure 4. C.O.V.E.R.T. Recon Approach

Again, consider COVERT as a basic plan for operationally stepping through the reconnaissance phase of a physical red team operation. Let's start by studying each of the six steps.

<u>Consume RoE</u>

By now, the RoE should be in hand and contain, at a minimum, enough information to begin planning reconnaissance. Remember, the RoE is a living document and may not contain much detail just yet. However, it should contain enough information to launch a recon mission of substance.

To get things rolling, the information consumed and analyzed should at least amount to the following:

- Target locations (addresses and GPS coordinates)

- Targeted people (specific individuals and/or employee roles)

- Google Earth photos

- Targeted controls

- Out-of-scope controls

- Operational objectives

- General idea of the complexity of TTPs

The team should not move onto the next phase until this basic information is acquired and understood. In fact, it might be helpful to copy and paste this information from the RoE into a separate internal document that can be easily reviewed by the team. It could prove useful as a quick reference cheat sheet as the team goes through the COVERT process.

Since the recon team will soon be going onsite to conduct recon, authorization from the client must be obtained from everyone but the recon team and a few client stakeholders. Employees, civilians, security forces, and bystanders may think the recon team's actions look like anything from a burglary in process, to terrorists in action, to a swat team raid, to trespassers, to plain old creepy dudes. As a result, each recon team operator must carry a federal/state ID and an authorization letter spelling out the reasoning behind his/her actions in the event they are compromised. As an aside, another name for an authorization letter is a "Get Out of Jail Free" card.

The authorization letter should contain the red team operator's employer, company address, employer bio, and after-hours contact info that will prove employment and vouch for the services being offered. At a minimum, the letter must state the nature of business that the team is conducting and after-hours client contacts (at least three) who can be called at any hour to substantiate the recon team's legitimacy in the event the team is apprehended by law enforcement, employees, or other security forces.

As stated earlier, an ID and the authorization letter must be carried by each and every recon team member. Please see the following link for an editable template of an authorization letter.

For an editable Authorization Letter (aka: Get Out of Jail Free Card) template in Microsoft Word, please visit the following URL:

https://www.redteamsecuritytraining.com/physical-red-teaming-book

<u>Obtain Targets</u>

The RoE will have some basic information about targets, but certainly not exhaustive amounts. In this step of the COVERT process, we want to dive deeper into where and who our targets are and what might be around them.

Open-source Intelligence (OSINT)

At this point, we will have physical addresses, GPS coordinates, maybe some aerial photos from Google Earth, and perhaps a handful of staff names we are targeting. Using addresses, we should be searching open sources like Google to find images and information relevant to our targets, etc. This leg of recon is what most folks refer to as Open-source Intelligence (OSINT). OSINT is data collected from publicly available sources to be used in an intelligence context.

When the team becomes aware of a target name, say Jeremiah Talamantes, our goal is to turn that name from a 'John Doe' into a persona. What do I mean by that? Searching my name in Google will turn up other people that share my name, who you'll find you can quickly dismiss. However, you may find some other interesting information about my career, schools I've attended, businesses I own, and other personal interests. Slowly my name evolves from merely a name into a full-fledged persona. This intel may serve to be valuable in targeting me later, or it may not. But the principal points I am making here are to personify the raw information we have in the RoE using OSINT tactics and turning it into something of value to the operation.

Here are some OSINT resources you can use to personify an individual, a company, and/or its facility:

- Google Images and Google Earth

- Google Dorking

(https://www.exploit-db.com/google-hacking-database)

- Twitter, LinkedIn, Crunchbase, Indeed, Monster

- Recon-NG, Maltego

Verify Goals

Given the information obtained during the previous steps in the COVERT process, now is the time to set reconnaissance goals and ensure they align with the RoE as a whole. One way to do this is by reviewing the RoE's objectives and targeted security controls. Then ask yourself, what is the operation's overall objective? What are the client's most critical assets? What controls are we testing, who are the likely bad actors, and how sophisticated are they? Answering these fundamental questions will enable the development of reconnaissance goals that, in the end, will feed the latter phases of the operation.

Let's take a look at an example. Let's assume our client is a critical infrastructure power company that owns substation facilities that have small huts providing network connections into its SCADA and internal network. The client has nearly 100 of these small housing structures in substations spread across a wide geographic footprint. As the red team, we suspect their physical security posture might not be adequate and likely pose significant threat by likely bad actors through these substation structures.

During this step of the COVERT process, we need to set recon goals that enable us to find out whether our suspicions of these small huts is correct. We might set a goal to covertly recon a few substations in an attempt to learn what cameras, motion detectors, and personnel are present. In every physical red team operation, there will likely be several recon goals just like this all serving different purposes but unified in support of the RoE and the operation in its entirety.

Make a list of the recon goals and share them with the team. Feel free to use and adapt the sample recon goals shown in Figure 5.

Recon Goals					
GOALS	Goal	Plan	Est. Vulnerability	Threat	Bad Actor
1	Monitor personnel traffic at substation A to H. Identify physical security controls (cameras, motion detection, locks, RFID)	[PROVIDE DETAILS ABOUT HOW THE TEAM WILL EXECUTE THIS GOAL]	Inadequate perimeter security	Moderate to Significant Service Disruption	Nation-state, moderately sophisticated
2					

Figure 5. Sample Recon Goals List.

Estimate Resources

Among the most essential parts of estimating resources are time, travel, and tools. Estimating resources is a breeze if you follow those simple steps.

Time

Time consists of both operation time and red team operator time. To begin, we need to understand the client's needs and relevant deadlines. Of course, that will vary from client to client.

It is in everyone's best interest, however, to carve out as close to a finalized project timeline as possible as early as possible. It's especially critical when it comes to both the recon and the execution phase since this usually involves travel. Poorly coordinated travel is one of the biggest reasons operations fail.

Some red teams will conduct their initial recon and then immediately go into the execution phase during the same trip to the site. The REDTEAMOPSEC method separates these two occurrences for this very reason. As a result, each red team operation will require a recon team and an execution team, both involving separate trips to the site. Oftentimes, the recon team consists of the very same execution team, minus an operator or two.

When considering how long it may take onsite, I always ensure there are at least three days for onsite recon. This gives time to capture intel during the day and night on more than one occasion as opposed to a single day and night. Some recon goals may be accomplished by one operator, while the rest of the recon team achieves a different goal. Good planning will help the team use the time onsite more efficiently.

Travel

Once a project timeline is established, it now becomes essential to estimate team resources. How many red team operators will it take for recon? Which operator will do what? How long do we need to be onsite? Some of this information will become evident once recon goals are defined. Most recon missions can be done with three operators, occasionally only two. It's a good rule of thumb to adopt at least three operators for every recon mission.

Travel plans should be made early for at least three operators. In order for the team to get acclimated, the travel plan should include at least one full travel day. The team could use the extra time on a travel day to review the recon goal list one more time and prep tools or other equipment.

Tools

By now, the recon goals list is widely known, a project timeline is there, and the recon team may be assembled. A big part of the necessary tools can be sourced from the recon goals list. The team should take the time to ensure each of the tools are in working order and make efforts to purchase or make any others.

Let's not forget that clothing is also part of this step as well. Terrain and weather considerations may force the team to bulk up and make new purchases as a result.

Ready Team

This step assumes the team has well-defined recon goals, is fully equipped, and has arrived onsite. You could consider this step a mini version of step five in the REDTEAMOPSEC method called, Execute Staging. This is the staging phase, one of the last steps before the team switches from passive recon to active recon.

To help certify readiness, here is a quick checklist of must-haves:

- Every operator is carrying an authorization letter and a federal or state identification

- Every operator is carrying the necessary gear

- Every operator's gear has been checked and is in working order

- Every operator has a means of effectively communicating situation reports (SITREP) to the team during execution

- Every operator has an assigned recon goal and knows their role in accomplishing that goal

- Every operator knows what signifies mission success for each recon goal

- Every operator knows what to do in the event of a compromise by an employee, a bystander, law enforcement, or security force

- Every operator knows where the rally point is located

<u>Team Executes</u>

Go Red Team! This is the most exciting part of running recon missions. Tensions are high, the adrenaline is flowing, the team puts their planning into action and actively moves into position. This marks the point at which an operation could go sideways if compromised. It is critical that the team stays on task and follows the plan closely.

Truth be told, no amount of planning will guarantee that the mission will go off perfectly. Expect hiccups along the way and take time to play out what you would do in potential and unfortunate scenarios.

Some say reconnaissance is far more of an art than a science, and, to some degree, that's not entirely untrue. I have found success in operationalizing it with the COVERT method, as we've seen here, and the Recon C^3 Method as you'll see next.

Recon C^3 Method™

I developed the Recon C^3 method to enable recon teams to stay on task and approach reconnaissance missions uniformly and clearly. Refer to the illustration in Figure 6.

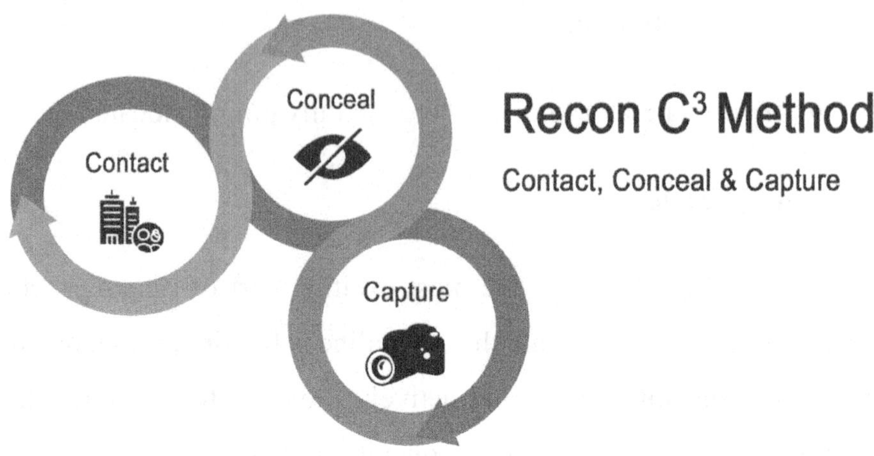

Figure 6. Recon C3 Method

The Recon C³ method includes three critical execution phases of recon missions: **Contact, Conceal, and Capture**. Despite the simplicity of the C³ method, reconnaissance missions can and do sometimes go off the rails if not managed properly. As stated earlier, this method offers a quick and easily digestible way for recon teams to stay on course. So, let's take a moment to unpack each of these phases and examine a little further.

Contact

The primary step during the execution phase of a recon mission is to make contact with the target or targets. Contact can happen in several ways depending upon the objective. However, most missions start out through surveillance from afar. This could amount to the team watching the movement of people as they come and go from a targeted building. It could mean using Google Earth to capture aerial photos of the target. Making contact, in another example, could mean engaging in conversation with a targeted person with the goal of surreptitiously obtaining information from them.

Alternatively, making contact could mean using covert methods of entry to break into a building under the cover of night. Essentially, making contact is the first step in what we call active reconnaissance and marks an important delineation between recon planning and recon execution.

The contact phase is important for another primary reason. If the recon team is seen doing something suspicious by an onlooker or an employee, the recon mission could be compromised. Depending upon the circumstances, the entire operation could be compromised in a manner significant enough to warrant aborting it altogether. So, the introduction of this significant risk should not be taken lightly, and the team should proceed with caution.

We will discuss how teams should exercise caution during the contact phase later in this chapter.

Conceal

Next in the Recon C^3 method is the conceal phase. It is important to note this can mean many things depending upon the recon objective. However, in most scenarios red teams quite often hide their physical presence under the cover of darkness while taking photos and video of a target from a distance, for example. Yet in other situations, operators may make their presence known to human targets, but might be concealing discreet recording devices in order to capture recon intelligence of importance.

As I stated earlier, it is very common to conduct recon missions in hiding. This can happen from outside a building, incognito in front of people, and so on. In nearly all situations, clothing becomes among the most valuable concealment tools. Wearing camouflage outside an industrial building is no different, in principle, than wearing a business suit in a corporate environment. In both situations, the objective is to blend in without drawing attention. The same concept is applied to discreet tools, such as a pen camera or an eye-glasses camera. Thus, the concept of concealment often relates both to the red team operator herself and the tools used to acquire intel.

Capture

The capture phase is fairly straightforward and is the end goal in any recon mission. No, we're not capturing hostages. We are capturing information, by video and photo, that will allow us to analyze and make predictions about where there might be vulnerabilities. The information we capture almost always includes the following:

- Physical security controls (fences, barriers, cameras,

entrances)

- People (attire, traffic patterns, civilians vs. employees, roles)

- Places (surrounding businesses, cafes, restaurants, traffic)

- Terrain/Weather (urban, rural, sunny, snowy, desert, rocky)

Analysis performed at the capture phase feeds the rest of the REDTEAMOPSEC process and gives light into the operation's specifics such as: how, who, what, when, and where.

With that brief introduction of the C^3 methodology, let's strap in and dig into the heart of recon mission execution.

Executing Recon Missions

We have the simplicity of the Recon C^3 method to use as a high-level guide during execution. Worth mentioning is, each of the three Cs should be carried out from, at least, two physical vantage points, from afar and up close. Simple, right? I typically refer to this as long-range recon and short-range recon. So, as we step through each of the three Cs, I will divide the action steps into long-range and short-range categories. Let's get started.

Contact

Beginning with the first of the three Cs, let's examine the contact phase first.

<u>Long-Range</u>

The first, and arguably the easiest, contact to make is through the internet. Earlier we discussed using open-source intelligence (OSINT) sources to find information about a target. I'll outline several OSINT resources here to help achieve valuable long-range recon. You will conduct this leg of recon from your office.

Company Websites

Scouring your target's website is an obvious first step, and it usually pays off handsomely. Search engine optimization (SEO) experts advise companies to post unique and feature-rich information not only about their business but their culture. In a show of corporate transparency, interior photos, employee pictures, offices, hobbies, and even musical tastes are shared there.

For all the reasons stated, recon teams should scrape the company website in search of a range of topics. I suggest searching and documenting the following information:

- Location address(es)

- Employee names, email addresses, phone numbers

- Technologies used (Careers page)

- Social media accounts

- Photos of exterior and interior spaces

Advanced Google Search

Advanced Google searches are also called "dorks" by those who use them often. What's it all about? Well, it's a way to use Google's advanced search options to find more specific information about something. A dork is a string of text containing advanced search parameters that Google's search engine interprets and displays for you. It's common to use a Google dork to search for Microsoft Word files that contain certain keywords, such as your client's name, for example.

I recommend searching for Microsoft Office file types in conjunction with your target's URL. For a list of Google dork syntax and more, please see the following resource.

For a list of recommended Google Dorks, please visit this URL:

https://www.redteamsecuritytraining.com/physical-red-teaming-book

For lots more Google dorks, please visit: https://www.exploit-db.com/google-hacking-database

Google Images, Google Earth

I suggest searching your client's name in Google and then clicking the Images tab. I suggest using your client's name and location as search keywords. This search alone can provide some amazing intel.

Google Earth is a great resource for providing an aerial view of the target. There is not a single operation where we have not used Google Earth in some form or another.

Here is what you should be looking for:

- Entrances, parking lots, recon vantage points

- Surrounding structures and businesses

- Any images showing the interior

- Any visible security controls in place (internally, externally)

- Surrounding terrain

- Any metadata to determine the image's age

Job Portals

Job boards can give away a great deal of information about the company's security maturity. You can use sites like Indeed.com and Dice.com to gain information about the technologies used by the target. This would be especially useful if your physical red team operation has an element of technical penetration testing involved.

Here's what you should be looking for:

- Company infrastructure and job roles (Physical security role?)

- Tech stack (Windows versus Mac)

- Security posture/practices

- Contact information (phone, email, names)

LinkedIn

LinkedIn is a great resource to find information about individual targets. At the time of this writing, LinkedIn has an option to upgrade an account with more features. I suggest upgrading so you can view any LinkedIn member and so that members won't know you viewed their profile.

Here is what you should be looking for:

- Important staff members, roles, location, and responsibilities

- Staff's previously held jobs

- Peers

- Staff interests, education, and accomplishments

Facebook, Twitter, & Instagram

- Company culture, news, initiatives

- Employee manner of dress (casual, business)

- Emails, phone numbers, staff names

- Interior photos

- Job openings

<u>Short-Range</u>

Short-range contact is made by the recon team making their way to a physical location near, but not at, the target recon areas. This first position is called the staging area.

An ideal staging position should be taken up near the target recon area but far enough away to allow for the team to meet without arousing onlooker's suspicion. In rural and less dense areas with minimal coverage, the staging area may teeter between a short driving distance and walking distance. In more urban areas, the staging area is usually within walking distance and around corners or in alleys. The recon team should make a few passes around the selected staging area and recon areas to re-evaluate their viability. My personal recommendation is to use a dark-colored passenger van with ample space to stage from.

The staging area should be considered a safe zone, but at least one member should be cognizant of their presence to onlookers and move to another location should suspicions rise.

Once at the staging area, the team should meet and finalize details before taking up their next recon positions. This is often an opportune time to review recon goals.

Figure 7. Staging Area

Figure 7 depicts a large building, our fictional target, and illustrates one possible staging area behind several long rows of cold storage buildings. This staging area is ideal for its vehicle cover and less visible approach from the rear of the targeted building.

Recon of all locations should happen during the day and at night in an effort to capture the most comprehensive intelligence possible. Most often, the same staging area position can be used during both nighttime and daytime.

Conceal

At this step, the recon team must stage for concealment. Operators must adapt their clothing according to:

- Time of day

- Weather/terrain

- Recon goals

Time of day matters in rural areas where there is less lighting and less structural cover. At night, black tactical or dark camouflage may be better here. This type of clothing at any hour in an urban setting around people should be avoided unless the team feels highly confident they can avoid bystanders.

In areas of rough terrain, combat boots by 5.11 Tactical are essential. A wide array of ripstop clothing can be found in tactical and everyday fashion form as well.

Ultimately, recon goals are the largest predictor of clothing requirements. What is the information you're after and how will you get it? Sometimes it means walking right up to somebody and talking with them. Occasionally my team will engage in conversation with bystanders or even target employees. In every single one of these situations, we wear street clothes. So again, the goal will make the garb.

Naturally, team members use tools to capture recon intel. Sometimes they use discreet cameras hidden on their body while other times they use a video camera hidden in a shoulder bag. In every situation, the goal is to get close enough to make the determination that a vulnerability exists. Having video footage to study later helps tremendously in making that determination.

Figure 8. Bag containing hidden camera. Photo credit: Paul Szoldra/Tech Insider

Figure 8 depicts a member of my team showing his shoulder bag that was built to conceal a hidden camera inside. Concealment tools and tactics like this enable operators to capture recon intel in open areas without drawing unwanted attention.

Figure 9 shows another team member using an everyday laptop bag to conceal a portable RFID reader. As he made his way through the office, he managed to capture and later clone the RFID badges of individuals with high-level building privileges. These badges were later used to make entry into the facilities at night.

Figure 9. Bag conceals RFID cloner. Photo credit: Paul Szoldra/Tech Insider

How and what to conceal depends upon the recon goals defined earlier in this process. The options are endless and are limited only by your imagination and creativity. That said, I want to provide some concealment tactics and tools that my team uses on a regular basis.

Vehicle Hide

Your team should take photos of the location from a distance. My team almost always does this from the recon vehicle. Earlier I recommended a passenger van. Passenger vans have many windows providing multiple vantage points and space to move around.

A vehicle hide consists of window coverings that prevent outside light from coming in that could potentially expose the photographer while providing just enough space for the camera lens to poke through. You've seen these in spy movies, I'm sure. They can be fashioned with scissors, a black bed sheet and some double-sided tape. Cut the black sheet into sections large enough to cover all rear windows and use the double-sided tape to attach the top and bottom areas of the sheet to the interior. A hole can be cut into the sheet just big enough to poke the camera lens through. Vehicle hides help tremendously in avoiding detection.

Bag Hide

Figure 8 illustrates a great example of using a bag hide to conceal a camera inside. In this example, we cut out a nickel sized hole in a shoulder bag and poked the lens of a GoPro camera through it.

Phone Hide

If your coat or shirt has a breast pocket, simply put your

smartphone in video record mode and use it to capture video evidence. Most smartphones come equipped with a camera at the top of the device making it an easy way to record video from your shirt pocket without drawing attention.

Hand Hide

On several occasions, I have covertly cupped a GoPro camera in my hand as I made my way past a facility. These small form factor cameras make it super easy to take video, and quickly stash it in your pants pocket.

Pen Hide, Button Hide, Glasses Hide & Other Discreet Cameras

There are online marketplaces, such as Amazon.com, that are full of discreet cameras too numerous to be listed here. Some of the cameras I've used include pen cameras, button cameras, and eye-glass cameras. These cameras are generally useful only in good lighting and within close proximity to the target. Consider investing in at least three discreet cameras in various form factors. Each has their advantages and disadvantages and should be used in appropriate situations.

Capture

The capture phase marks the last step in the Recon C^3 method. To add clarity to this critical phase, please consider the following process diagram.

Figure 10. EDECE Process for Recon Capture

Figure 10 shows the five-step **EDECE Methodology** (pronounced "ED-eh-see") for the capture phase of recon missions. Let's unpack the EDECE method as we step through our recon mission.

Long-Range

Establish Rally Point

A rally point is a physical location where the recon team will go to once their recon mission is completed or aborted. In the case of multiple operators, there may be several rally points depending upon the location of the operators. In our example, there will be only one rally point.

The red team leader is almost always positioned inside the recon vehicle at the rally point. This is where she will command the team and communicate with the client throughout the mission. A good rally point is one that places the vehicle close enough to provide eyes-on surveillance of the target, yet far enough away to go unnoticed. Operators should be able to walk to the rally point without difficulty and without arousing suspicion.

Figure 11. Selected Rally Point

The red team leader must then communicate the location of the rally point to the rest of the team and to the client. Next, the team should get ready for deployment into the field.

Deploy Team

Before making the trip to deployment locations, an equipment check and a communications check should be performed. Radio communication is recommended for night deployments, while smartphones and Bluetooth earbuds are recommended for daytime deployments when around people. In daytime situations, it helps for the entire team to hop on a conference call.

Figure 12. Deployment Areas

Team mobilization is made toward each deployment area in numeric order. Operators should be deployed in a staggered timeframe with at least five to ten minutes in between. See Figure 12 for ideal deployment positions in our example. Deployment positions should be selected for their close proximity to the target while providing cover for the operator's exit from the vehicle. All operators should exit from the passenger doors as opposed to the van's rear doors. Rear door exits often look suspicious to onlookers.

Deployment positions should be selected so that their vantage points offer the most physical coverage of the target. Notice in Figure 12, each operator is deployed from opposite ends of the building. Immediately before exiting from the recon vehicle, the operators should:

- Have their authorization letter and ID present

- Perform a communications check (phone or radio)

- Perform a gear/tools check

- Switch their video capture gear to record

- Communicate to the client that recon is about to start

- Take a deep breath and try to relax

Short-Range

Engage Target

Each operator should have at least one hands-free discreet camera or body-worn camera in record mode to capture the target's exterior as they approach the target. The footage from these vantage points gives additional visibility and oftentimes helps uncover security controls not previously noticed from afar.

A second camera should be available and aimed at fulfilling their recon goal. From my experience, this second camera is typically a GoPro stashed away in a pocket, an iPhone, or a pen camera. You want a device you can quickly grab to capture intel and then stow it away as necessary.

In our example, let's refer to the operator deployed at location #1 as operator #1 and so on. Operator #1 should make her way along the designated route. Refer to the dotted pink line shown in Figure 13.

Figure 13. Operator #1 Recon Route

Operator #1's goals are to capture physical security controls installed to the rear of the building. The loading docks are of significant importance since very few access controls are present there. A large sweep of the rear with a hands-free camera will provide useful footage to review at a later date.

Operator #1 must maintain contact with the team and alert them of any onlookers taking an interest. She should also indicate whether the scene is clear to deploy operator #2.

Figure 14. Operator #2 Recon Route

Operator #2's route shown in Figure 14 has her hugging the front and west sides of the target. She will then travel along the sidewalk along the highway then veer north to capture the east side before reaching the rally point.

Capture Intel

Each operator will have one or more areas of interest that she will likely pause to observe and record. These recon goals will vary from operation to operation. That said, here is a viable list of recon goals the team in this example would likely try to capture.

Operator #1 (Refer to Figure 13)

- Loading dock doors, trucks, dock entrances

- External lights and their focal points

- Cameras, motion sensors, motion activated lights

- Barriers, fences, walls

- Entrances and signage

- Entrance locks

- Surrounding activity (foot traffic, vehicle traffic)

Operator #2 (Refer to Figure 14)

- Lobby, lobby doors, lobby security controls

- Lobby activity and personnel (receptionist)

- Office activity, employees, manner of dress, roles

- Vacant offices to the west

- West side office doors, locks, physical security controls

- External lights, internal lighting, and their focal points

- Cameras, motion sensors, motion activated lights

- Alternate entrances, locks and signage

- Surrounding activity (foot traffic, vehicle traffic)

Some of the goals here can be captured by simply walking past the area with a camera. Yet other goals, such as examining locks, often require the operator to stop and record. For these stop-and-record situations, I recommend faking a phone call while walking slowly near office windows and while you walk up to video record door locks. The more distracted you appear, the more likely you will not be called out by an onlooker.

Here are a few false scenarios you can use for stop-and-record situations:

- Distracted telephone conversation. Use this scenario to walk slowly and occasionally stop to capture recon. Be animated as you talk and use your hands to gesture.

- Walk, Stop, Text. Stop and position your camera to record the desired area while you pretend to type out a long text message.

- Enter the office and ask for directions. Ask directions to a nearby restaurant or café and capture footage of internal spaces. Be careful whenever approaching targets face-to-face this early in the operation.

- Mistaken delivery. Fake a pizza delivery to the office while you capture footage of internal spaces. This usually provides more time on the inside as employees hunt for the owner of the pizza. Select an independent pizza restaurant and be sure to dress appropriately. Again, be careful whenever approaching targets face-to-face this early in the operation.

These fake scenarios, or pretexts, can be adapted to fit many different recon goals. They can certainly become more aggressive depending upon the need to penetrate deeper into the facility.

As I have mentioned previously, a second tour of recon capture should be made during nighttime hours. Since most business offices are closed around the midnight hour, this second tour will have a slightly different focus.

Here are some common nighttime recon goals:

- Target activity (cleaning crew, overnight security), if any

- Surrounding activity (foot traffic, vehicle traffic)

- How well-lit is the target and where are the unlit areas?

- Locations of motion-activated lighting, if any

- Viable infiltration points based on these factors

The security posture of a facility can drop significantly simply by the time of day. As a direct result, most red team infiltrations are attempted during the wee hours and thus it become increasingly critical to estimate just how much the environment changes.

Exit to Rally Point

Things can sometimes go sideways during recon execution. To prepare for the unexpected a backstory, or pretext, is necessary. Each recon operation must have a plausible explanation in the event of a compromise. For believability, each operator must rehearse and become comfortable reciting their backstory with confidence.

Communication amongst the team is important, largely when a team member has been or is about to become compromised. The red team leader must make a snap decision in response. Usually, the red team leader will give the other team members the command to exit quickly and gracefully. This is done to minimize the likelihood of further compromise leading to complete mission failure. The uncompromised team members should then make their way to the pre-determined rally point unless directed otherwise.

In contrast, upon successful completion of the recon goals by the team, each operator should communicate to the red team leader accordingly. If the red team leader feels the recon goal has been met, she should give the command to exit to the rally point. For large recon coverage areas, there may be several rally points but, in most cases, there will be only one pre-determined rally point.

When the exit order has been given, the red team leader should stagger exfiltrate procedures so that the entire team is not seen leaving all at once. As a rule of thumb, the operator who faces the highest likelihood of being compromised should be exfiltrated first and so on. Each operator should follow the pre-determined exfil route as planned, unless conveyed otherwise by the red team leader.

Before leaving their position, each operator should perform a quick equipment and tools check to ensure nothing is left behind that might indicate their presence. If necessary, steps should be taken to cover any physical tracks. That said, our ground coverage at this point is considered light and non-intrusive when compared to the operational execution. We will discuss covering tracks in greater detail later in the REDTEAMOPSEC method.

The team should make their way to the rally point calmly and in the same manner as their initial approach. The team should enter the vehicle through the passenger door and move to the rear allowing any following team members to enter similarly. Once all members have returned to the recon vehicle, the team should perform a more detailed equipment/tools check to be sure nothing was left behind.

As the vehicle departs the rally point, all team members should immediately document their findings and observations while they are still fresh in memory. Some of my team's most telling observations were incapable of being caught on camera, so it's very important to have note-taking supplies, paper tablets, or laptops handy. By now, the red team leader should communicate to the client that the team has completed the mission and has departed the target location.

A recon debrief session should be held once the team is able to meet and discuss their observations, photos, and video. Each operator should provide an overview of their recon goals and go over their related recon findings in detail with the team. The red team leader should compile all of the recon footage, photos, and notes as each operator presents their findings. This data will feed into the next phase in the REDTEAMOPSEC method as deep-dive analysis takes place and preparations are made for progressing through the operation.

The next phase of the REDTEAMOPSEC methodology is called Direct Preparations and is central to transforming raw data from the recon mission into valuable information.

For hands-on Physical Red Team Training, please visit:

https://www.redteamsecuritytraining.com/physical-red-teaming-book

[CHAPTER 4]

DIRECT PREPARATIONS

Successful physical red team operations would not be possible without a strong game plan. Goals sourced from the RoE and raw recon intelligence power the operation and are vital to the preparations phase of the REDTEAMOPSEC methodology.

In this chapter, we will cover the preparations and analysis necessary to further the red team operation. We will analyze the recon intel and align the action steps to follow in support of the RoE. This step is critical and paramount to the success of the operation as a result of the decisions made here. Some of the decisions made at this step will be made as a team. However, most of the heavy lifting in this phase will fall upon the shoulders of the red team leader.

Here are the primary areas of concentration for this chapter:

- Recon Intel Review

- Vulnerability Analysis

- Additional Needs & Resource Planning

- Operational Plan Development

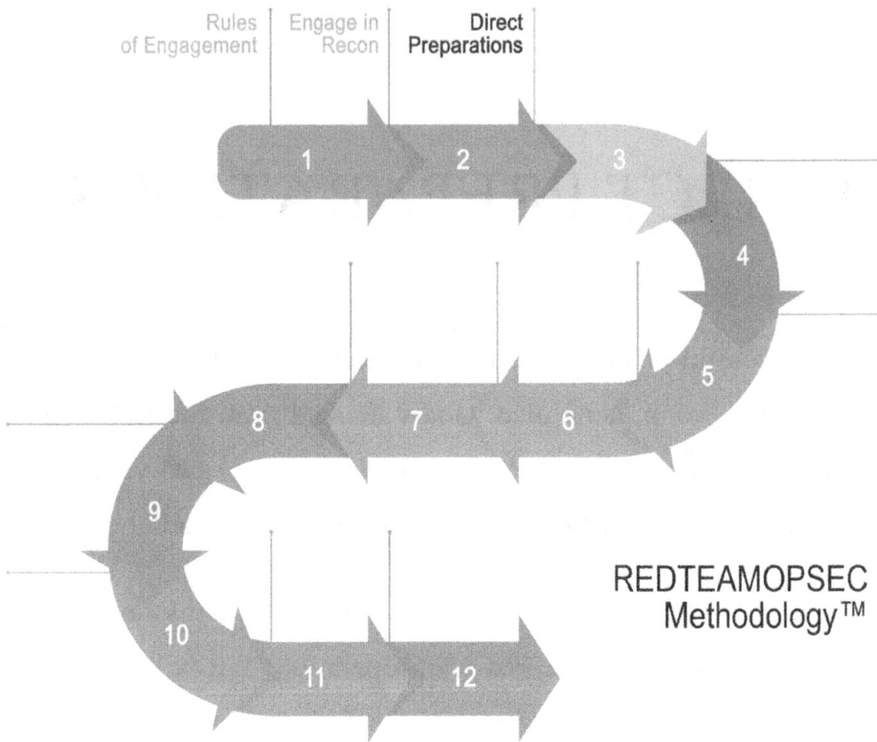

Figure 15. Direct Preparations Phase

Recon Intel Review

Immediately following a reconnaissance mission, the recon team should meet to debrief. To prepare for the meeting, the red team leader should gather all notes, footage, and photos from each of the team members. Once all recon artifacts have been collected, the red team leader should begin creating a timeline of events. Oftentimes, the timeline of events is a client-facing document that summarizes the mission's beginning, end, and milestones in between. I've found that clients like to use these timelines to better understand why their current security controls did not detect or trigger an incident when reviewing security camera footage, guard tours, etc.

When reviewing recon artifacts, each recon member should provide their recollection of events including times, captured footage/photos, and observations. This verbal walkthrough should be detailed, in chronological order, and align with previously defined recon goals. Every artifact (video, photo) should be reviewed by the team. Any observations unable to be captured by video or photo should be entered into record by the red team leader. By now, the red team leader should have a more complete timeline of events to be shared with the client. If so desired, the red team leader could upload the timeline to the document repository for client review.

Having discussed the observations and outcomes of each recon goal together, the team must analyze the results.

Vulnerability Analysis

Vulnerability analysis is a critical part of the preparations step. At this point, the team must analyze each outcome to answer the following:

- Given the recon observations, does the team believe there is a vulnerability present?

- Is there a security control in place to defend against exploitation of this vulnerability? If so, how sophisticated is it?

- Is the likely bad actor more sophisticated and/or better resourced than the in-place security control?

Recon Goals				Results	
GOALS		Goal	Plan	Observations	Vulnerability	Go-Forward?
	1	Surveil loading dock area, search for insecure doors and possible entry ways. Also take note of any security cameras and motion lights.	Deploy from area #1, approach from rear on foot and walk along alley way. Wear blue-collar clothing and fake a phone call while pausing in front of entrances. Capture footage and take every opportunity to video door locks up close.	Loading dock traffic is moderate by day, non-existent by night. No cameras visible, have motion lights. No RFID badge entry, only door locks. External doors have ADA levers and weather stripping below.	No motion alarms. No security cams. Motion lights can be bypassed on east side. ADA lever handles could be bypassed with under-the-door-tool and air wedge. Infiltration to happen at night.	Yes
	2	""	""	""	""	""

Figure 16. Finalized Recon Goals Table

Please see the Results columns in Figure 16. This shows the completed version of the Recon Goals table presented earlier in this book. Some columns have been hidden for space. The red team leader should complete each recon goal with a summary of the observations, vulnerability, and decision to move forward with testing or not.

As stated earlier, the team must analyze each recon goal's

observations to determine if there is a vulnerability present and if testing is applicable. In some cases, recon intel may indicate adequate protection or better. In most cases, this determination is usually not so apparent. The decision to go forward with testing is one that should be made as a team. Most importantly, if a vulnerability is present, would the likely bad actor be sophisticated enough to exploit it? How well resourced would he need to be? Is this a vulnerability that is worth testing? These are all important questions to be considered. Remember, vulnerabilities should be tested in a commensurate fashion with the level of sophistication of would-be perpetrators.

Additional Needs & Resource Planning

As the vulnerability analysis portion comes to a close, the team has decided which security controls pose issues significant enough to be tested. The team should also have a clear idea of how to test those issues. As a result of all this, there will likely be changes. They might find a need for additional tools and maybe even more team members. It is important at this step to ensure the team makes, purchases, or sets aside these extra necessities. This is a pivotal point in the process as we prepare for execution down the road.

From my years of experience, this phase almost always involves changes to the RoE. So now would be a good time to revisit the RoE and update as necessary. Any RoE changes must be shared and approved by the client before moving forward.

All in all, a lot of this planning can occur without involving the client. But this marks an important point at which operation dates and times should be finalized. This will need to be coordinated with the client and must be done as far in advance as possible.

To prepare for immediate next steps in the REDTEAMOPSEC phase, such as Trigger Mobilization, a second staging site should be chosen at this time. If you recall from Chapter 4, the staging site is where the team suits up, tests gear, and makes final preparations minutes before deploying into the field. The next staging site should be ideal for specific execution purposes. You may find that your previously used staging site during reconnaissance is still an ideal spot. However, understand that most operations take place at night and could involve wearing dark clothing, carrying strange tools, or behaving suspiciously. Be sure the next staging site is conducive to those activities.

Here is a short list of preparation items that should be covered here:

- Finalize TTP strategies and adapt plans and resources appropriately

- Acquire additional tools and equipment before moving forward, if necessary

- Add additional red teamers to the operation, if necessary

- Finalize operational dates and times with client

- Coordinate travel plans for the execution of the operation

- Determine the ideal Staging Site (see "Execute Staging")

- Update the RoE and obtain approval from the client

Operational Plan Development

Before going any further, I will link to an editable Microsoft Word template of an Operational Plan below.

For an editable Operational Plan template in Microsoft Word, please visit the following URL:

https://www.redteamsecuritytraining.com/physical-red-teaming-book

I will provide a brief outline of the same Operational Plan to follow. However, I strongly encourage readers to visit the link instead, as the document will be updated and improved periodically.

The list to follow is an example of a high-level outline of an operational plan. I recommend using the existing components at a bare minimum. You will need to modify as necessary. Again, this is a client-facing document that is presented and discussed following the Engage Reconnaissance phase. The plan is subject to change. Therefore, client stakeholders must have complete visibility and be required to review and approve any salient changes as they happen.

Plan development is where most of the effort will be concentrated during this phase. You'll notice the operational plan provides the client with some of the same information as the RoE, but with more meat, particularly number 11, titled Operation & TTPs. Operation & TTPs aligns perfectly with the REDTEAMOPSEC methodology and aims to summarize to the client what will happen at each point. Items A to K are what we will be covering in detail in the chapters to follow, so don't be alarmed if you don't know their meaning right now.

Operational Plan Outline

Here are the major components of the Operational Plan.

1. Client Name

2. Client Contacts

3. Project Contacts

4. Red Team Members & Roles

5. Target Location Address(es)

6. GPS Coordinates (Each location)

7. Photos of Location(s)

8. Operation Dates & Times

9. Operation Objective(s)

10. Target Control(s)

11. Operation & TTPs

    a. Reconnaissance

    b. Preparation

    c. Mobilization

    d. Staging

    e. Assess & Acclimate

    f. Maneuver Operations

    g. Strike

    h. Penetration & Control

    i. Execute Operational Orders (OPORD)

    j. Evacuate, Evade, & Cover

    k. Collect & Exfil

12. Out-of-Scope Control(s)

13. Out-of-scope TTP(s)

14. Damage Causing TTP(s)

15. Additional Notes

16. Document Change History Table

    a.   Change Description & Date

    b.   Client Change Review/Approval Signature & Date

For an editable Operational Plan template in Microsoft Word, please visit the following URL:

https://www.redteamsecuritytraining.com/physical-red-teaming-book

[CHAPTER 5]

TRIGGER MOBILIZATION

Many in the physical red teaming industry often overlook the importance of proper team mobilization to the staging site or target site. When is the right time to move out? What time of day makes the best time to infiltrate a facility? Where does the team meet? Do they roll out all together? All of these questions and more will be answered as we examine a better way to mobilize the team for execution purposes.

In the technical penetration testing industry, going from reconnaissance to direct port scanning, for example, marks an important milestone in the penetration test. From that point on, we consider the penetration test itself to have gone from passive mode to active mode. Being in active mode signifies a much more direct assault on the target and thus raises the risk of possibly being caught or compromised significantly. In the physical red teaming industry, we recognize this milestone similarly and refer to it as the **Execution Phase.** Trigger mobilization and the collective phases to follow in the REDTEAMOPSEC methodology are loosely referred to as the

Execution Phase.

Let's look deeper into the ever-important aspect of REDTEAMOPSEC we call Trigger Mobilization.

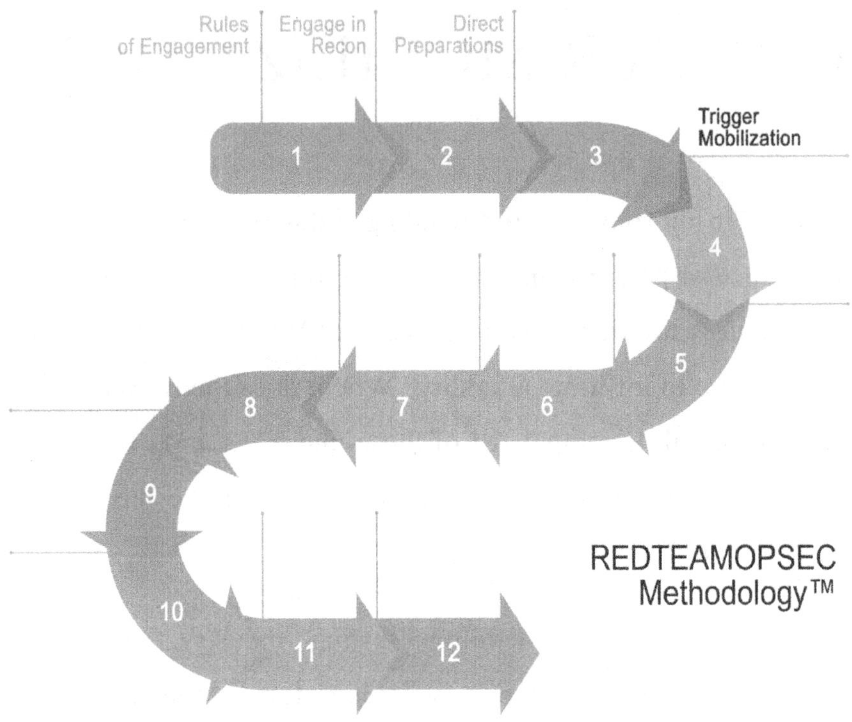

Figure 17. Trigger Mobilization Phase

The goal of Trigger Mobilization is to control the movement of red team operators to the target in a controlled and orderly fashion so as not to compromise the mission with their presence. To clarify, the team will mobilize to a staging site near the target site rather than at the target site proper. As I mentioned in Chapters 4 and 5, staging sites allow teams to suit up and test gear while being far enough away from the target proper to go unnoticed yet close enough for the team to deploy in minutes.

In short, Trigger Mobilization is a planning phase that offers us orderly team movement, a staging area for preparation, and quick team deployment.

Here are the important aspects of the Trigger Mobilization phase:

- Staging, Deployment, and Rally Point Selection

- Mobilization

Staging, Deployment, and Rally Point Selection

Figure 18. Staging Site Example

Staging Site

Recall from an earlier chapter where we selected a staging area for carrying out reconnaissance (Figure 18). The recon staging site allowed the team to review recon goals, suit up, test gear, and move into deployment point positions quickly. In our example in Figure 18, this site was chosen for its close proximity to the target and its line-of-sight cover from unwanted onlookers. Essentially, the same principles apply when selecting a staging site during this phase but with a slight twist.

From my experience, after the reconnaissance phase is complete, the team generally has a much better understanding of the target site's physical environment and its flow of people, traffic, nuances, etc. Thus, the previously selected site may not be ideal for execution. In fact, simply having stepped foot on the grounds and having obtained a ground-level perspective first-hand offers team members a chance to feel out the area.

During a recent operation, I recall picking up a very strong sense of tension in the air as my team descended upon the target, and my teammates felt it too. Perhaps it was the giant ominous spotlights blasting the grounds from the guard stations or maybe this location was "too hot," as we like to say. Something in our gut told us we should do things differently and stage somewhere else.

All in all, a lot can be learned from the initial recon mission. At a minimum, this staging site must wholly support execution purposes and a close proximity to deployment points. Thus, a different staging site is often selected for this reason. To make staging site selection easier, take the following factors into account when making the new selection.

Staging site consideration factors:

- Site selection supports execution goals specifically

- Consider lighting conditions (usually performed at night)

- Close proximity to deployment point(s)

- Trust your instincts if the site seems too hot

Deployment Point

Recall from Chapter 4, "Engage in Reconnaissance," where we discussed deployment points and their importance. Deployment points are areas on the map that signify where team members make their final descent upon the target. See Figure 19 for the sample deployment sites we chose in Chapter 4.

Deploying into the field marks an important milestone in the overall REDTEAMOPSEC methodology. What is unique about the Trigger Mobilization phase when compared to recon deployment is that the deployment planning steps aim to enable our entry into the facility. For this reason, there are a few differences in how one should go about selecting deployment points.

Figure 19. Sample Deployment Sites

Deployment site considerations factors:

- On the fringe of being too close to the target

- Consider security camera range, guards, motion lights, etc.

- Consider lighting conditions (too illuminated?)

- Does the site support quick deployment (e.g. exit from vehicle)

- Close proximity to desired entry points (e.g. doors, windows, fence, roof, fire escape, etc.)

Rally Point

Figure 20. Sample Rally Point

See Figure 20 for the sample rally point we chose in our example in a previous chapter. The rally point is where the team will assemble outside the facility and exit the location, usually by vehicle. So once a red team leader deploys her operators in the field, she will likely drive to the rally point and command the team from there. You can think of the red team leader as the getaway driver and the rally point as the getaway spot.

The same criteria that goes into the selection of a rally point is similar to the criteria that goes into selecting a deployment point but with a few differences.

Rally point consideration factors:

- Supports red team leader staying parked for duration of mission

- Considers proximity of personnel from target and bystanders

- Within two-way radio range of the red team

- Supports quick team exfil (e.g. pick up team)

- Considers lighting conditions (too illuminated?)

- What walking route will operators take to reach it? Do their clothes or belongings look especially suspicious?

Mobilization

A successful red teaming mission depends on coordination taking place at many levels. Among the many things central to success is team mobilization. Team mobilization boils down to the controlled movement of red team operators to a target in an efficient, orderly fashion. If effectively completed, the team will arrive intact, on time, and prepared without compromising the mission with their presence.

There have been a handful of occasions where I have experienced missions going sideways as a result of poor team mobilization. Thankfully, they didn't result in total mission failure, but they easily could have. It is important to know when and where the team needs to be extra cautious. Please have a look at the list below pointing out the stages that pose the greatest risk during mobilization. To paint a clearer picture, I will list them in order of risk.

Periods of high-risk during team mobilization:

1. Team member deployment (risk increases for multiple deployments)

2. Waiting for the team at the rally point

3. Team members leaving the target en route to the rally point

Movement to Staging Site

Movement to the staging site is usually done by vehicle. Optimally, this should be done using a single vehicle, like a passenger van. Passenger van windows allow the team to have eyes peeled in several directions and also support space for larger teams when needed. Traveling in one vehicle will aid in preventing unwanted attention.

Shown below is a quick list of steps during mobilization:

- Circle the staging site

- Reaffirm staging site still meets expectations

- Confirm team is ready for staging

- Approach staging site

- Execute staging procedures

Movement at this phase almost always occurs during the late hours of the night. The dark provides several advantages to the bad guys. The security posture of a facility changes drastically simply by the time of day. Again, this is something the bad guys know, and it is something that they use to their advantage. This is among one of the primary reasons that nearly every physical red team operation I have carried out has been at night. Simply put, it provides a more realistic test. But the cover of night can be blown if improper use of flashlights, for example, give away a team's location through poor light discipline.

On the topic of light discipline, the driver should maintain appropriate light discipline when it comes to the vehicle's headlights, brake lights, and interior lights. Drivers should kill all external lights and mute or turn off interior lighting. Internal light sources emanating from cell phones, laptops, and such must also be kept to a minimum.

Red team operators should maintain appropriate light discipline when outside the vehicle as well. Red lights and low lumen headlamps are ideal here. An important note regarding infrared let use: If the team is using night vision to spot infrared cameras or merely as a visual aid, their equipment will give off a signal visible to others on the same spectrum. Good infrared light discipline means using these tools where appropriate and in limited fashion. It is important to note that light discipline will soon become even more serious when the team reaches their deployment sites and beyond.

Once the vehicle has landed at the staging site, the team should suit up and prepare quickly but thoroughly. Final equipment checks, radio communications tests, and clothing changes should occur at this time.

[CHAPTER 6]

EXECUTE STAGING

The team has mobilized to the selected staging site by van. It's very late at night and pitch-black outside. Situated inside the vehicle, movements are hurried, red teamers are gearing up, chatter is high, adrenaline is flowing, and last-minute equipment checks are happening all around. Welcome to the Execute Staging phase.

This phase is filled with an ambience of controlled chaos and rightly so. This is one of the last opportunities for red team operators to make critical strategy modifications, test equipment, fix issues, confirm the RoE goals, and more. There are many reasons why this is considered the end of the line before heading out into battle on foot.

Here is a quick list of major steps for this phase:

- Confirm RoE goals

- Suit up

- Test equipment

- Check comms

- Deploy

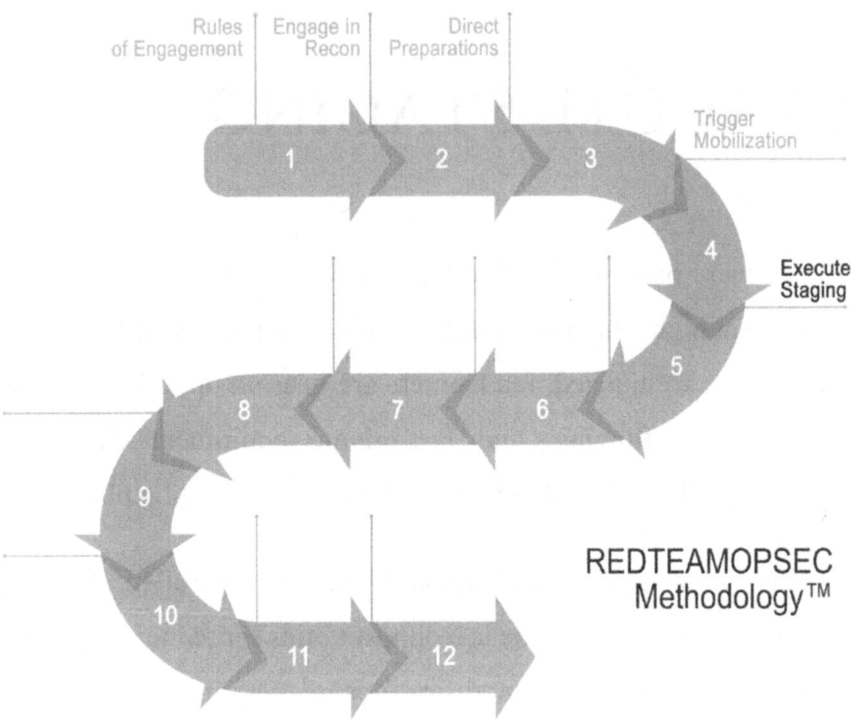

Figure 21. Execute Staging Phase

Confirm RoE Goals

This step is usually short and sweet but extremely important. The red team leader must confirm each red team operator knows their goal and how to successfully reach it as it relates to the RoE. Likewise, each operator must know out-of-scope TTPs, the rally point location, etc. A full review of the RoE is usually not necessary, but the red team leader should be satisfied with each operator's verbal confirmation.

Suit Up

You need to look the part. Legwork and decisions performed previously should enable operators to suit up appropriately. Whether the situation calls for black tactical gear, a suit and tie, or street clothes, the team needs to don suitable clothing for the mission. In Chapter 4, we presented a few helpful tips for choosing the right attire.

Consider the following when suiting up:

- Carry compartments

- Use of carry bags

- Time of day

- Weather/terrain

- Execution goals

Time of day matters in rural areas where there is less lighting and less structural cover. At night, black tactical or dark camouflage may be better. However, this type of clothing at any hour in an urban setting around people should be avoided unless the team feels exceptionally confident they can avoid bystanders.

In areas of rough terrain, combat boots by 5.11 Tactical are essential. A wide array of ripstop clothing can be found in tactical and everyday fashion form as well.

Mission goals are the largest predictor of clothing requirements. What is the team after and how will they get it? Sometimes it means walking right up to somebody and talking with them. Sometimes it means engaging in conversation with bystanders or even employees. The goal will make the garb.

In most of my missions, I've had to carry several tools along to complete the job, ranging from a laptop to a pick set to an awkward under-the-door tool. Clothing should suit your needs and have enough carry compartments to keep your hands free. As a general rule, hands should be kept free and tools should be stored in compartments as often as possible. I recommend a tactical vest equipped with MOLLE webbing. MOLLE is an acronym for Modular Lightweight Load-carrying Equipment.

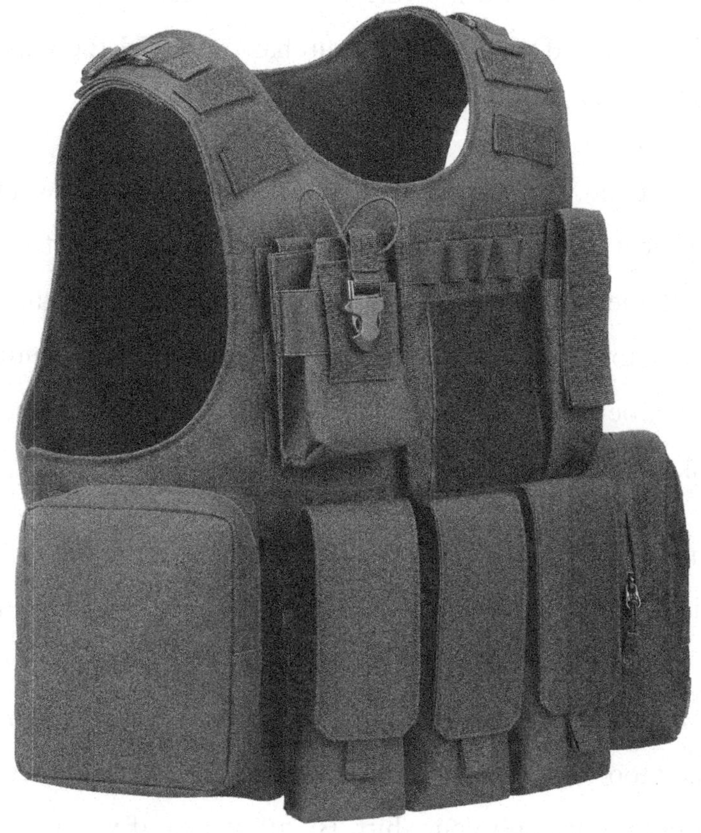

Figure 22. Tactical MOLLE Vest

The vest shown in Figure 22, or a variation of it, is ideal for carrying equipment while keeping hands free. The MOLLE webbing system provides an array of attachable pouches and compartments for any job. Many of the tactical vests on the market are used by law enforcement and military. As a result, they will come equipped with gun holsters, magazine storage, and other compartments that are not relevant to the mission. I recommend buying a basic MOLLE vest and purchasing MOLLE compartments separately to suit your needs.

To reiterate, be aware of the risk of onlookers when wearing any tactical gear. You can guarantee bystanders will report you to the police if they see you donning "suspicious" clothing like this in public. Before suiting up in any tactical clothing and venturing out in public, be sure the risk of detection by others is minimal or that the risk of detection is worth it.

Sometimes you need to conceal a tool that's too big to fit inside a tactical vest or a carry bag, like an under-the-door tool. Instead of using a MOLLE vest, you can substitute your pants and shirt and forgo any carry accessories. For example, you can hide an under-the-door tool on your person by situating it down one leg of your pants and up through your shirt. By all means, if you can afford to forgo any carry accessories, do so.

For equipment that is too large to fit into your tactical vest, consider using small tactical backpacks or slings. Figure 23 shows a tactical sling bag I use in the field to store larger pieces of equipment. The sling bag makes it quick to remove, open up, grab what I need, and mount it again with ease.

Figure 23. Small Tactical Bag

This tactical sling bag has MOLLE webbing on the front, side, and shoulder sling making it an ideal alternative or addition to a tactical vest. Tactical laptop bags and other larger form bags and backpacks complete with MOLLE webbing are also available.

I want to give a word of caution. Tactical bags, slings, and backpacks increase your physical footprint as you move about the location. This can make it difficult to climb a fence, squeeze in between obstacles, and navigate through offices. I could pause for a moment and tell a story about how my team member stealthily cleared an office only to knock down a big pile of paperwork with the butt of his bulky backpack, but that would just embarrass him. Sorry, Kurt.

The key to suiting up appropriately is to pack lightly, be aware of your physical footprint, and when you're in the thick of it, make every attempt to avoid leaving a trace ("a tell") of your presence.

Test Equipment

This one goes without saying, but when you're in the throes of staging, there can be a tendency to overlook things. In an earlier chapter, I recommended using a list of equipment for each mission, called a Load Out List. Oftentimes, it's the responsibility of one of our red team operators to use the Load Out List to ensure designated equipment has in fact made its way to the staging site and has been pre-checked beforehand.

Using a Load Out List during the staging phase will serve as a reminder to pack all the necessary equipment into bags prior to moving on to the following stage. It also helps ensure the equipment has been recently tested and that this round of testing should produce good results.

For an editable Load Out List template in Microsoft Word, please visit the following URL:

https://www.redteamsecuritytraining.com/physical-red-teaming-book

There is a long list of potential tools a red team operator could use here. It would be impossible to cover them all, but here are some last-minute reminders:

- Check the charge on batteries (cameras, NVG, radios, phones)

- All cameras, video and still, have adequate storage

- Take test shots and/or footage to ensure functionality

- Multipart gear has all necessary components (under-door-tool, lock pick sets)

- Double check each operator has all of their gear

- Secure tools inside carry bag and vest compartments

Check Comms

Communicating with fellow operators is vital to an operation. Using cell phones or two-way radios are both really great ways to communicate with red team operators during an engagement. We will dive deeper into how to manage communication during operations later in this book.

Figure 24. Two-way radios with earpiece accessory

My team uses two-way radios during most of our operations.

Radios generally facilitate cleaner communication without a lot of background noise and unnecessary "chatter" you sometimes get by using cell phones. Figure 24 shows an example of a model my team uses as well as the earpiece accessory. The push-to-talk button located on the earpiece can be placed either near your neck or down a sleeve for easy access.

Whether you prefer radios or cell phones for comms, it's important to communicate efficiently, clearly, and precisely. The team should use procedure words (prowords) for all radio communication. Prowords are often used by the military and no doubt you've heard these before. Some examples include, ROGER, LOUD AND CLEAR, COPY THAT, OVER, and so on.

Radio Prowords

Here is a list of prowords the team should use:

Operator's Proword List	
Proword	**Meaning**
Radio Check	Can you hear me okay?
Loud and Clear	Response to Radio Check. Your signal is good.
Roger	Message received and understood
Copy/Copy That	I understand you
Out	End of transmission – no reply is expected

Over	End of transmission and a reply is expected
Say Again	Repeat your last transmission
I Spell...	A word to follow will be spelled phonetically
Go Ahead	Ready to receive the transmission
Wilco	Yes, I will comply.

Figure 25. Operator's proword list

Before choosing a communication medium, be sure to know the pros and cons of choosing one over the other. Have a look at Figure 26 for a non-scientific evaluation on the topic.

Radios vs. Cell Phones

Radios vs. Cell Phones		
	Radios	**Cell Phones**
Hands-free	Yes (Earpiece)	Yes (Bluetooth)
Range	Limited by range	Excellent! Not limited by range
Reliability	Varies a lot	Excellent
Form-factor	Can be bulkier than phones	Varies, but usually less bulky
Clarity	Good. When within range	Very good
Efficiency	Excellent! Less chatter	Very poor
Weather	Very good	Very poor. Can't get wet
Durability	Great	Very poor
Usability	Excellent	Excellent

Figure 26. Radios vs. cell phones

Whether you've elected to use radios over cell phones, a comms check must happen for every red teamer. If the situation allows, comms checks should always be performed outside the earshot of other operators. Usually, this process occurs just outside the staging vehicle.

When my team uses cell phones, we often put in our Bluetooth earpieces and jump on a conference call. Perform a comms check by simply announcing yourself to the other callers.

Part of the comms check should be checking your radio battery or your cell phone battery life. Both devices should have LED lights and/or onscreen displays indicating as such. Always pack extra, already-charged radio batteries to solve low-battery issues on the fly. I can almost guarantee low battery issues will happen, no matter how well you prepare.

Low battery issues for cell phones can be solved with portable battery chargers like the one pictured in Figure 27. A universal charger, with a compact profile, capable of servicing Android and iPhone smartphones is ideal. This piece of equipment is a must-have for more than this reason alone and should become part of your everyday carry.

Figure 27. Universal Portable Cell Phone Charger

Deploy

The red team leader gives the "go forward" command once she feels satisfied prior staging steps have been completed to satisfaction. This phase is more of an attestation to the team's readiness to move out than a detailed series of actions.

Attestation

Just prior to departing from the vehicle, the red team leader must:

- Ensure all operators have adequate ID and an authorization letter on hand

- Ensure all operators know their mission goals and how to achieve them

- Have all the necessary equipment tested and in working order to achieve mission goals

- Have comms activated, tested, and in working order

- Have evidence capture devices (video camera) in working order and set to record

- Ensure operators know where the rally point is

- Re-check that all operators have adequate ID and authorization letter on hand

Once the red team leader can attest to the team's readiness, she can give the go forward command to deploy.

Formation

Upon exit from the vehicle, it may become necessary for the team to commence movement in formation. This is especially true when there are 25 yards or more to the perimeter. I find formation particularly beneficial in wide open terrain with little to no structural cover. Moving in formation will help a team facilitate orderly movement while reducing the risk of compromise.

Column Formation

In column formation, the team approaches in a forward movement keeping a single-file line. Figure 28 illustrates an example of column formation.

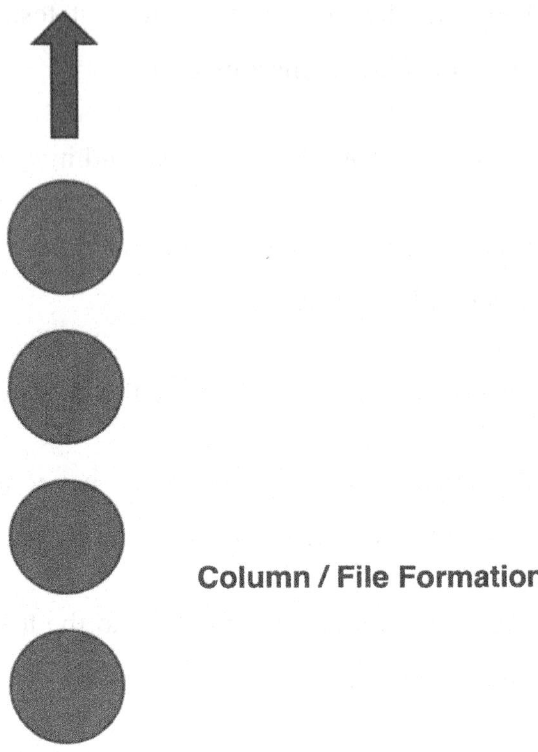

Figure 28. Column Formation

Because the team's forward movement is done in a tightly narrowed column, this formation is ideal for navigating across thick/bushy terrain. However, I find this tactic the most useful when it becomes necessary to conceal the team's presence. The narrow forward-facing presence is what makes this possible. One of the drawbacks is that the first person in the column (point person) is solely responsible for spotting visual threats.

<u>Line Formation</u>

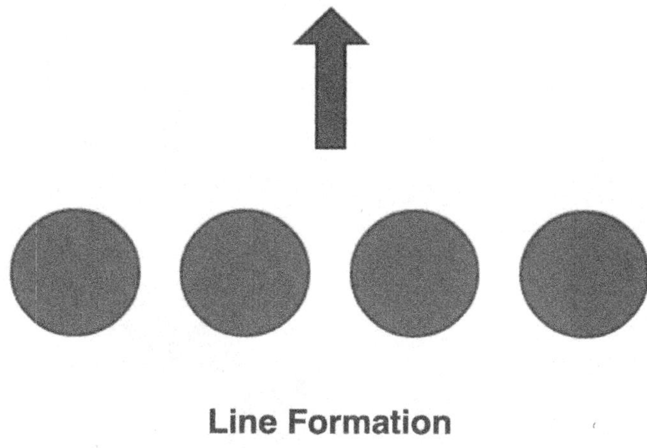

Figure 29. Line Formation

Unlike column formation, red team operators make their approach side by side. One of the biggest advantages is having more eyes facing forward to be able to scan for potential threats. By increasing the distance between operators carefully and dividing the forward motion into lanes, operators can scan a large swath of terrain easily.

An obvious drawback to this formation is that the team will take up more space on the ground and be more susceptible to identification. However, my team uses line formation when there is ample ground available, there is at least some cover, and the need to scan a broad area for potential threats is high.

<u>Vee Formation</u>

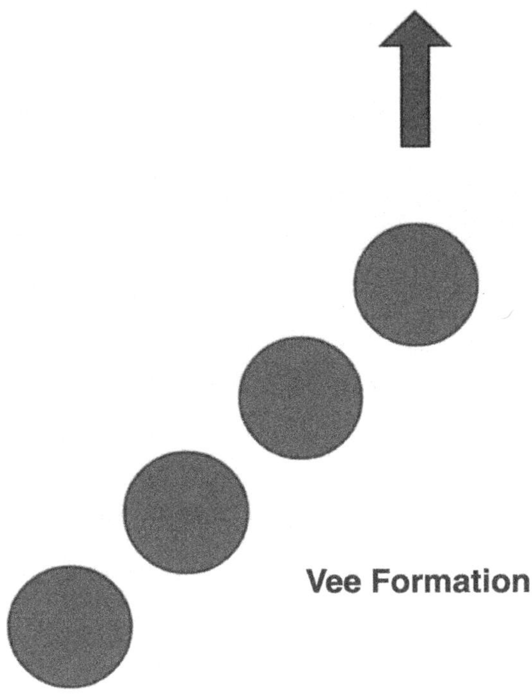

Figure 30. Vee Formation

Somewhere in between column and line formations, roughly speaking, is the vee formation. Here you can experience the benefit of multiple eyes forward along with the benefit of a narrow formation for covert purposes. Because this is a narrow form, it might be possible for the team to fall in a column line at a moment's notice, if necessary.

Vee formation is one of the most versatile and the one I recommend, when in doubt. True, there are many other military formations to adopt here. However, from my experience there is no real need to complicate the process any further.

For hands-on Physical Red Team Training, please visit:

https://www.redteamsecuritytraining.com/physical-red-teaming-book

[CHAPTER 7]

ASSESS & ACCLIMATE

Under the cover of night, the red team leader gives the go-forward signal, and the red team exits the vehicle. As the red team leader drives away from the staging point, a powerful feeling of excitement, nervousness, and a little bit of fear swirls in the bellies of the team. Shaking off some of the butterflies, operators instantly recall their training, hours of mission planning, and a dash of bravado begins to take hold.

Reams of video and images of reconnaissance intel captured earlier in the engagement fill the team's minds as they situate themselves with the once familiar surrounding air, smells, terrain, buildings, lighting, weather, and security controls. The familiar fit of the ripstop clothing pressing against the skin and the added weight of their gear reminds them of the need to breathe easy and maintain a low profile.

The radio starts to buzz and crackle as the other team members instinctively fall into vee formation, take a knee, and pause. Then the radio begins to crackle into life. Is that a new security camera? Is there a light on inside? Is it busier than normal? Are the guards on a tour? How is this place now different from the Engage in Recon phase and what are we going to do about it? You are now entering into the Assess & Acclimate phase.

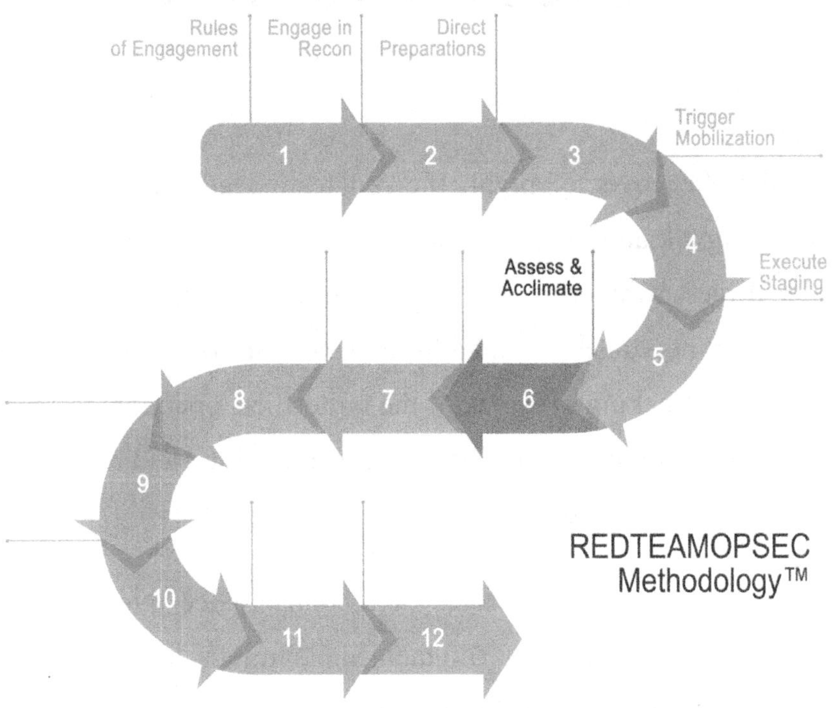

Figure 31. Assess & Acclimate Phase

The team is preparing to embark on the mission by foot, but before maneuvering operations for an offensive strike against the facility, they must first identify salient changes at the target. In other words, what important differences are there now and do these differences negatively affect the team's ability to reach mission success?

An example of a salient change might be the presence of additional security cameras, fencing, or motion sensors. Because some amount of time can pass between the reconnaissance phase (Engage in Recon) and this phase, the environment is subject to change without notice. In some cases, the differences could be pretty drastic. Thus, it becomes extremely important to assess the environment for changes before blindly advancing to the next stage.

Assess

Unless you have somehow uncovered additional intel about recent security improvements inside the targeted facility, you can only assess what you can see in front of you. So the facility's outside may look the same, but security improvements, like biometric scanners, placed internally can throw a wrench into any plan. Lucky for you though, most physical security improvements are placed outside and are typically visible.

You see, unlike the principle of defense in depth common to the information security industry, most physical security departments have yet to adopt that strategy. So instead we see a lot of physical security controls implemented in only one or two layers, almost always outside the building (fencing, cameras). That said, do not rest on your laurels. Plan for change.

It's worth noting, a great deal of assessing happens initially when boots hit the ground, there's no doubt about that. But as the team advances, the process of assessing and acclimating will be ever present. It's very difficult to plan for the unexpected. After all, that's why it's called the unexpected! There will be disappointing surprises. The best you can do is try to be as prepared and rehearsed as possible.

Now as you might've already guessed, the assessment process is one that needs to happen quickly. To make things even trickier, not all changes to the target environment will affect each team member in reaching their goals. The team must assess environmental changes as a team and communicate them. Then it becomes the responsibility of each team member to know how the change may affect their capability to reach their mission goals and adapt.

Here's a quick breakdown of how this should occur:

1. The team performs an eyes-on assessment for any changes in the environment

2. Operators communicate any changes over the radio

3. Individual operators evaluate how these changes affect their capability for reaching their own mission goals

4. Individual operators announce over the radio how these changes affect their capabilities

Consider Figure 32. This is an example of one team member's mission goal to gain unauthorized access through the target's loading dock.

GOALS		Mission Goals						Recon Results		
		Goal	Plan	Mission Success	Est. Vulnerability	Threat	Bad Actor	Observations	Vulnerability	Go-Forward?
	1	Gain unauthorized access through the loading dock. Capture video evidence and leave a business card.	Deploy from area #1, approach from rear on foot and walk along alley way. Wear black tactical gear and use under-the-door tool and air wedge to gain access. Capture video evidence and leave business card.	Gaining access via door exploit, capture evidence, leave business card and exit without detection	Inadequate perimeter security	Moderate to Significant Service Disruption	Local to Regional Bad Actor. Minimally sophisticated.	Loading dock traffic is moderate by day, non-existent by night. No cameras visible, have motion lights. No RFID badge entry, only door locks. External doors have ADA levers and weather stripping below.	No motion alarms. No security cams. Motion lights can be bypassed on east side. ADA lever handles could be bypassed with under-the-door-tool and air wedge. Infiltration to happen at night.	Yes
	2									

Figure 32. Sample Mission Goal

Now, it is not likely that each operator will have a printout of their mission goals with them to reference. So clearly, each and every operator must be able to recall every aspect of their goals from memory.

Operators can perform an assessment by recalling the data from the Recon Results columns along with the Goal, Plan, and Mission Success columns (Figure 32). Any recent security changes, upgrades or downgrades, that deviate from gathered recon intelligence have the potential to change the magnitude of the vulnerability or possibly even make the vulnerability disappear.

With guidance from the red team leader and the rest of the team, the team's course of action includes:

- Abort

- Advance as planned

- Acclimate and advance

Acclimate

When something changes at the target environment and it is believed to have an impact on the success of the mission, the team must react accordingly. It isn't all that often that a facility's security posture changes dramatically over a short amount of time. But minor changes can add up, and the team needs to acclimate as a result.

Let's assume the team identified a change in the environment and communicated the issue over the radio. Refer to Figure 32. Let's also assume we confirmed the presence of motion sensing lights at the loading dock where there were not any observed earlier. First of all, kudos to our target for taking steps to increase their perimeter security. After additional examination and collaboration, the team believes they can avoid tripping the sensor by moving slowly far and away from the sensor's focal point. The team also believes that a tripped sensor will not likely attract attention due to the somewhat concealed location of the loading dock.

This is a simple example of how the team can acclimate to changes in the environment. Here are a few things to consider when weighing different strategies:

- Does this strategy put the mission overall in unnecessary jeopardy?

- Does this strategy comply with the rules of engagement?

- Does this hinder or prevent other red teamers from completing their goals?

- Will this hinder or prevent me from completing any other mission goals?

- Do I have red team leader approval to proceed?

Assessing and acclimating to a changing environment is more of an art than a science. With practice, this will become easier.

For hands-on Physical Red Team Training, please visit:

https://www.redteamsecuritytraining.com/physical-red-teaming-book

[CHAPTER 8]

MANEUVER OPERATIONS

The team has identified minor security changes in the environment compared to the recon intel collected earlier; some additional security lighting near the loading dock for starters. The team quickly assesses the situation and decides on a mitigation plan.

It's very dark outside and there may be more hidden surprises. Crouching low to keep a low profile, they fall into a tighter vee formation and advance carefully toward the target. Several minutes have passed, but it only seems like seconds. It feels like things are happening too fast. But it always feels this way.

Senses are intensified so much that your ears start to buzz. Eyes dart from here to there, and heads swivel side to side as the target's external façade slowly grows more visible. Is that shadow a newly added security camera? Do I hear a car approaching? Did I forget to pack anything? What if there is an alarm on the door? It seems far too bright. Should we re-route our approach?

Breathing is noticeably more taxing now, but the equipment isn't that heavy. The team's earpieces chirp, and a static-filled radio announcement temporarily breaks their focus. It's a familiar voice. The red team leader announces over the radio that she's reached the rally point and wants a situational report (SITREP). The team prepares to break into two groups and strike.

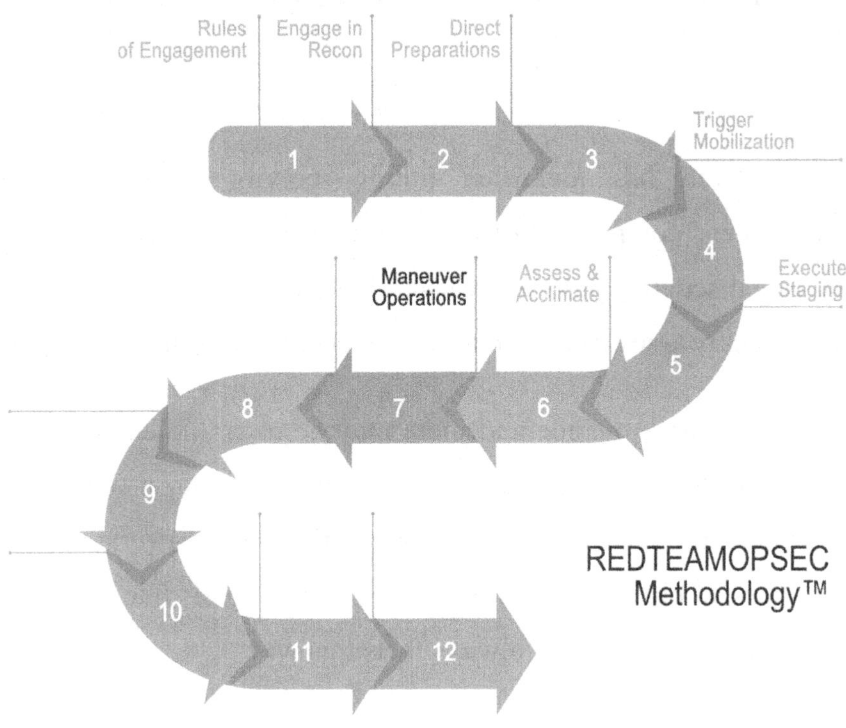

Figure 33. Maneuver Operations Phase

It is fundamental to any physical red team operation to move swiftly, securely, and confidently to and around a target. In fact, if at any point during the operation the team is going to be compromised, it will likely happen here. This is because the team's actions are often exposed to the public eye and at the same time close enough to be scrutinized by the target facility's cameras and security personnel.

The Maneuver Operations phase has a few key components. Here they are:

- Environmental Conditions

- Settlement

- Observation

- Cover & Concealment

- Movement

Environmental Conditions

It may seem like most physical red team operations happen in urban settings in moderate climates, but that's not entirely true. Think about data centers, supply warehouses, outposts, and remote offices. Data centers, these days, aren't always located in bustling metropolitan areas. Instead, remote locations are sought out for their mild climate, stable power, and discreet locale.

The important thing to note is that locations and environmental conditions will vary, and it's essential to know how to prepare for some of the challenges they can pose.

Types

To begin, here are a few types of environmental conditions we will cover:

- Dry & Arid

- Wet & Rainy

- Cold & Snowy

Please note this is a non-scientific approach, but I believe it to be brief and sound enough to cover most of what is important here. The following guidance assumes a rural or semi-rural target location.

<u>Dry & Arid</u>

In parts of the US where it is dry and desert-like, you will usually find some difficulty traversing the ground. Dry, rocky terrain produces uneven earth and plenty of opportunity to slip and slow movement. Furthermore, the threat of exposure is elevated due to a general lack of cover and concealment. At best, scattered knee-high to waist-high bushes provide the most consistent cover.

There are several types of dry and arid environments, the majority of which lie scattered in the western half of the U.S. and focused in the southwestern U.S. A quick study of Köppen climate types overlaid on the U.S. map will provide much more insight here (see Figure 34). Labels ranging from "continental warm summer," like we see in the upper Midwest, to "hot desert" in Arizona, California, and parts of Texas can help prepare teams tremendously.

Köppen climate types of the United States

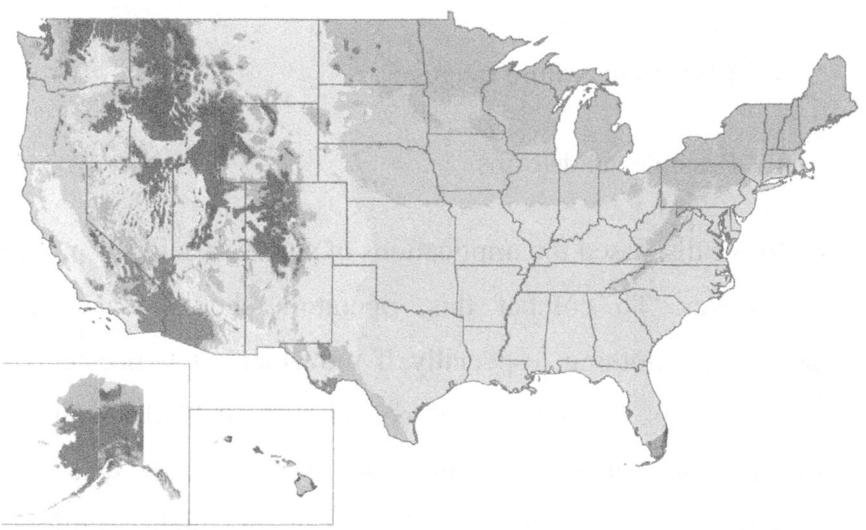

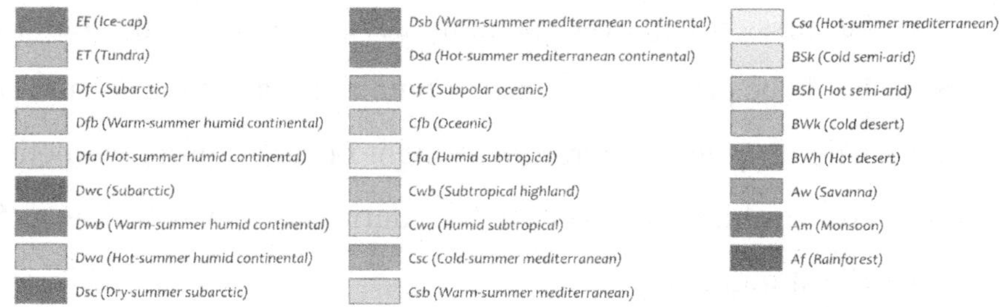

Figure 34. Climate Types of the US

Preparing for dry and arid environmental conditions, one should consider the following:

- Sparse vegetation

- Intense heat and sunlight

- Wide temperature ranges

- Possible sandstorms

We've discussed the implications of sparse vegetation providing little cover. To combat this, operators should make use of camouflage clothing, especially if movement is expected during daylight hours. When in doubt, consider desert MARPAT (marine pattern) camouflage. Clothing should be breathable so as not to make the wearer uncomfortable in the heat and intense sunlight.

If the mission calls for spending several hours in the field, the team should be prepared for the hot humid day and the cool and cold night. The drop in temperature at night occurs rapidly, sometimes going from as high as 130° F (55° C) during the day to as low as 50° F (10° C) at night.

Footwear should have ankle support to help prevent rolled ankles and be rugged enough to withstand tough terrain. Be aware of other hazards such as snakes that often live in these dry environments. Carry plenty of water to stay hydrated.

<u>Wet & Rainy</u>

According to Figure 34, humid subtropical climate makes up a great deal of the southeastern United States, and cool, rainy weather makes up the northwestern corner. Weather conditions are fairly moderate in most of these areas, aside from the potential for rain. Wide temperature ranges are not as common unless you traverse into some of the bordering Mediterranean areas of the northwest.

The likelihood of rain and windchill contributes to the potential for hypothermia. In these conditions, rain clothing, warm undergarments, and waterproof equipment are essential. Terrain in most of these areas is often not overly rugged and uneven. Still, combat boots or hiking boots should be worn. Shrubs and thorns are probable in most of these areas. Ripstop pants aid in preventing scratched legs.

For movement during the day, woodland MARPAT is recommended. To account for the presence of trees and grasses, this pattern is green and much darker than desert MARPAT.

<u>Cold & Snowy</u>

Physical red teaming in the winter is treacherous, especially in the northern United States. Keeping warm is crucial. Do not let the objective of a mission jeopardize one's health and safety. My company, RedTeam Security, is headquartered in St. Paul, Minnesota. We experience temperatures of -20° F and windchills of

-45° F in the winter. In cold conditions, warmth and safety must be primary. Snow boots, woolen socks, long underwear, thick undergarments, heavy outerwear, gloves, and ear/face protection are a must.

Cold doesn't always mean snow. Depending upon the location, woodland MARPAT might still be a good camouflage pattern to wear. However, where there is ample snow on the ground, white snow camo is preferred. There are many pattern variations. In the upper Midwest, where big game hunting is popular, many sporting goods stores will have them in stock.

Snow boots and clothing alone will add noticeable weight, making it hard to move around with any kind of speed. Snow can drift significantly high, making normally flat ground hard to traverse. Move slowly and carefully; you never really know how deep the snow is and what is underneath.

Of course, the weight of these environmental conditions ultimately depends on where the target location resides. Clearly, we don't need to wear camo if the target is located on State Street in downtown Chicago.

Settlement

How one goes about coordinating the movement of an operation not only depends on the environmental conditions, but also the setting. As I mentioned earlier, not all operations play out in urban areas. My company performs physical red team operations in various settlements, and urban areas make up the least common setting. This is why most of the guidance in this book assumes a non-urban setting.

Let's go over two distinct and common settings, urban and rural, and the nuances of each.

Rural

Rural settings aren't limited to flattened farmland. They also include small towns and communities. Targets in rural settlements have characteristics about them that make them unique. First, let's take a moment and list some of the most notable characteristics, and then I will go on to detail a little bit about them.

- Abundant acreage

- Bystanders on high-alert

- Physical security patterns

In my experience, targets located in rural areas generally have ample acreage surrounding their facilities. This often means more ground to cover before reaching a target's perimeter. The downside is a greater window of exposure on approach, yet the upside is the team can make use of the acreage to setup mini-staging sites. More ground can also mean more cover and concealment opportunities.

the added landscape becomes the backdrop, it's important to choose the right camouflage clothing for concealment. There are certainly pluses and minuses to extra acreage. All in all, plan to traverse more ground and plan for a longer time in the field. Movement at nighttime is ideal compared to daytime.

Wide open spaces mean there is less chance of being visually compromised by passersby. I've often noticed that urban dwellers tend to disregard suspicious behavior slightly more than rural residents. I recall an incident where my team was literally pulled over by a rural resident who wanted to know who we were and what we were doing in his small town at 3 a.m. His expression proved that he didn't buy the backstory we gave him, and he proceeded to follow our vehicle for the next 20 minutes. In rural areas, it is important to maintain a low profile and travel lightly. The presence of vehicle headlights in a normally quiet area may be enough to put a bystander on high alert.

I recall another engagement where the project stakeholders warned us about residents near our target who were very liberal about carrying firearms. Understand that any resident, rural or urban, could be carrying a firearm and potentially mistake your intentions as threatening. It's important to maintain sound situational awareness. Again, maneuvering an operation is usually better when done under the cover of darkness.

Physical security controls for rural targets tend to follow certain patterns. The most common security controls include motion activated lights, chain link fencing, and residential door locks. We find these low-grade security controls the only thing protecting some of the most "critical" facilities. Facility owners often feel the rural environment reduces the likelihood of something bad happening. We must teach them this is not true!

Urban

Let me point out some of the ways populated areas are different from rural when it comes to physical red teaming. This is probably one of the most telling from my experience. It seems that urban bystanders have a much higher tolerance for reacting to suspicious behavior. In one of the more densely populated areas, another operator and I were spotted making a covert approach completely decked out in black tactical gear late one night. We were carrying two ladders and a few tactical bags, yet the group of bystanders didn't give us a second look. We changed our entry point to be on the safe side, but we successfully compromised that location without incident. We've been seen by onlookers a few other times, but none as visually damning as this incident.

The next biggest difference is how drastic the environmental characteristics change from day to night. By day, a given target may be bustling with businesspeople, and by night, it may change into a less desirable location. During one engagement, my team and I literally walked into a small homeless camp with about four or five people. The approach had been surveilled and cleared during the daytime, but nightfall brought a completely different vibe. Thinking on our toes, we pretended to be building security and they picked up and left without confrontation.

Red teamers should never bet on city folk always being more tolerant of suspicious behavior. I suggest toning things down and using tactical clothing only where reasonable. Substitute all black tactical clothes with dark colored street clothes. Perhaps a mixture of black and gray colors. Use a dark colored laptop bag instead of a black tactical bag. Put a pair of over-the-ear headphones around your neck to be a little more convincing. By being sensible, you can easily alter your appearance to a less aggressive one.

Urban locations generally mean less acreage to traverse along with more opportunities for cover and concealment. Use this to your advantage and map out a route that places the team along a movement path with less chance for visual identification. Dumpsters and walled off trash inlets are common in cities and can provide good coverage. Alleyways, side streets, parking lots, and parking ramps should be sought out for their benefit of concealment. Figure 35 shows how a team might use adjacent structures to shield their deployment, staging, and rally points to avoid unwanted attention.

Figure 35. Sample Deployment, Staging, Rally Points

Finally, a thorough recon mission of the target location during both daytime and nighttime hours is mandatory. Urban areas change the most during a 24-hour period. If the initial recon is not fresh, within the previous week or so, the Execute Staging and Assess & Acclimate phases will help mitigate any deviation from collected intel.

Observation

From the Execute Staging phase and beyond, the team will be on high alert for individuals and security controls that could potentially spoil the engagement (security guards, staff, bystanders, law enforcement). That is a substantial amount of time to be extra cautious with plenty of opportunity to overlook someone or something.

In urban areas, one usually has less visibility beyond 200 yards due to buildings and other structures. This forces a quicker situational response time from operators. Generally speaking, the opposite is true when in rural areas. Either way, it is important to know how to cope in these scenarios, and I hope this short section will be of assistance.

Situational Awareness

The textbook definition of situational awareness is the perception of environmental elements and events with respect to time or space, the comprehension of their meaning, and the projection of their future status. The term is fairly new, but the concept isn't. Situational awareness tactics are employed by military groups, survivalists, and personal safety advocates, and there have been many adaptations to fit other industries.

For a Hollywood-style dramatization of this, watch the 2009 Sherlock Holmes movie or the 2002 Bourne Identity movie. In short, situational awareness is being keenly hyperaware of environmental elements, how they may (potentially) interrelate to create an incident, and how best to respond accordingly. Entire books have been written on the subject and science, but here are some tips to improve situational awareness.

- Baseline the environment. What is the typical amount of traffic? How busy is it here? What has changed recently? What is the status quo?

- Tune out distractions. Know what to look for, listen for, and smell. Sensory overload can and will happen. Focus on the path forward, the target, and any obstacles preventing mission success. Tune out the rest.

- Prepare to be caught off guard. Mentally prepare for this and you'll be more apt to respond positively. Know your backstory if someone approaches you. Rehearse what you would say in your head. Look for conceal and cover opportunities two or three steps ahead.

- Be spatially aware. Are your teammates directly behind you or 10 yards behind you? What does your position look like from all sides? Are you casting a long shadow? How does your position look from the top floor of your target? How many steps to the target and how long will it take to get there?

- Continually evaluate. What new obstacles are presented as you advance on the target? What risk level do these obstacles bring? What should I change to mitigate the risk?

- Shun complacency. Don't be a cowboy when things seem to be running smoothly. As I said earlier, prepare to be caught off guard. If your heart rate is at a cool 70 bpm, you've probably become too complacent. Be cautious and be hyperaware.

OODA Loop

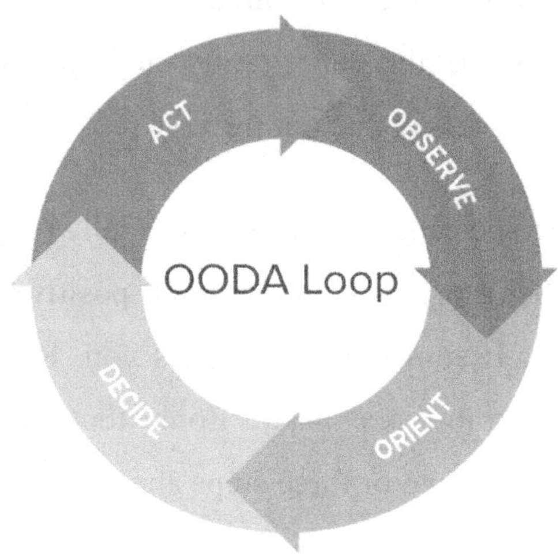

Figure 36. OODA Loop

OODA is an acronym for Observe – Orient – Decide – Act. The OODA Loop is a decision-making model originally developed by military strategist John Boyd to enable quick and sound decision making by troops in the field. According to its creator, decisions are made in a recurring loop, or cycle, which is where part of its name comes from. Allow me to unpack the OODA loop a little more clearly.

<u>Observe</u>

Really, this amounts to taking in the environment and all of its dimensions. Sights, sounds, smells, spatially, situationally, contextually, and emotionally, for example. Observing is the process of taking in all of this raw data, unfiltered, with the intent of transforming it into useful situational information.

Visual observation is extremely vital to this step. Is something out of place? Is someone approaching? Is that a camera? Generally, movement occurs under the cover of night where there are obvious visibility issues. Images become less sharp. Colors are hard to discern. Depth perception is off. And night blindness can set in.

Here are a few tips to help with visual observation in dark areas:

- Use night vision goggles (NVG) whenever possible

- Allow time for eyes to adjust before moving into a dark area – up to 20 minutes (if time permits)

- In a pinch, shut your eyes briefly when going from a lit to a dark environment

- Conduct staging procedures in a red-light area vs. a white-light area

- Do not stare at objects too long – look to all sides of targeted objects or pathways

- Scan the horizon slowly from right to left for about 3 seconds before going the opposite direction

<u>Orient</u>

To sidestep for a moment, OODA loop purists maintain there are five factors that go into the Orient step: Cultural Traditions, Analysis & Synthesis, Previous Experience, and Genetic Heritage. In my opinion, some of these factors do not easily translate toward red teaming as much as others. I believe one's own previous experiences and ability to analyze and synthesize situations are what are most adaptable to physical red teaming.

If an OODA loop were an artificial intelligence engine, the Observe step would be the data model and the Orient step would be where the model is trained. The Orient step is, no doubt, where most of the important magic happens.

Now it comes time to transform the raw data into information. The Orient step involves analyzing and contextualizing observational data collected, using previous experiences and new information, to recognize and orient against potential threats. In a physical red team engagement, being spotted or drawing unwanted attention can be detrimental and requires constant attention. To do this properly, it is important to understand the "norms," or baseline profile, of the area. What is the expected level of traffic for the area? How bright is the area this time of night? Any deviation from baseline expectations should trigger a reaction to reorient.

<u>Decide & Act</u>

In our previous artificial intelligence example, if the Orient step is where the data model is trained, the Decide step is where the ideal result of the analysis is provided. In other words, what is the optimal course of action available to mitigate or prevent the given situation? This process usually happens fast and feels more like a gut reaction than anything.

The decision of how to react to an unexpected approaching car may be to seek cover and avoid being seen. But where? The dumpster in front of you or the structure behind you? Decisions will vary depending upon the factors that go into the Orient step; predominately the decision-maker's own experiences and her ability to situationally analyze.

Once the decision-maker has acted, it becomes important to turn the loop quickly. The quicker one can turn their loop, the odds increase in their favor in evading a sticky situation.

Cover & Concealment

Cover and concealment provide natural or man-made protection from the unwanted attention of staff, guards, bystanders, etc. Trees, bushes, high grass, ravines, and the sort provide natural concealment, preventing operators from being discovered. Buildings, small structures, dumpsters, vehicles, and parking ramps are just a few examples of man-made cover.

Some of the smallest depressions in the ground can help prevent you from being spotted in open areas with little natural cover. I'll get into traversing this kind of environment in the next section. Avoid skylining your silhouette on a hilltop or other high point. This is hard to avoid in areas of flat terrain. So, be spatially aware of your profile everywhere you go, as well as the shadow you cast.

Figure 37. My team using cover during an actual operation

In my experience, man-made cover is often the most used during engagements. Since facilities, large and small, are our targets, there are often other surrounding structures to use as cover. Plan paths of movement with these structures in mind to prevent being spotted.

High grass, bushes, and trees provide the most utilized natural concealment. Concealment will not likely completely shield your profile but may be good enough in the interim. Where natural concealment is the only option, add layers of concealment such as camouflage clothing. As discussed previously, this simply means matching your wardrobe to the color and lighting of your environment. This also means using dark or camouflaged equipment. For example, don't use a flashlight that is colored safety orange.

In any environment, light, noise, and movement discipline all contribute to concealment. Light discipline is defined by the controlled use of lights to avoid drawing unwanted attention. Flashlights, headlamps switched on all the time, leaving vehicle headlights on, and smoking in the open are all examples of poor light discipline. The same light discipline holds true for infrared-enabled devices (IR) as well. Night vision goggles (NVG) emit a light, much like a spotlight, visible by other NVGs. Be aware of this and utilize NVGs only when necessary. I spoke about this earlier in the book, but use lights sparingly and where applicable. Situate the light only upon the subject of interest.

Binoculars and photo and video cameras all have various size lenses. Similar to light discipline is controlling the glare or shine these lenses may give off. In the right conditions, binocular lenses act like a signaling mirror and reflect light back toward the source. If the light source is coming from the target, this could be enough to grab the attention of security personnel and put the team in jeopardy.

Noise discipline is another difficult animal to control. We take for granted how much our voices carry and how loud dry leaves crackle under our feet. My team and I have never been too concerned with noise discipline outside in urban areas since noise pollution is common. However, once inside a facility, the audible environment will likely change. Sometimes it becomes quieter, sometimes it becomes noisier. The key thing to remember is to keep the volume of movement lower than the facility's natural ambience.

Movement

Up until now, movement from designated staging and deployment sites, as evident from this chapter, has been far more than just bodies moving from one position to another. There are many things to consider before skipping into a building. But for now, let's concentrate on proper movement for both teams and individuals in the field. In addition to walking, there are a few individual movement techniques to use throughout an engagement – rushing, high crawl, and low crawl.

Figure 38. Low Crawl

Low Crawl

The low crawl, often referred to as an Army crawl, provides an operator with the smallest silhouette during movement. This is often used in wide-open rural areas where there is very little cover or opportunities for concealment. When my team has resorted to low crawling, which isn't often, it is only performed for a short distance. This usually happens when within close proximity to guards on tour, security cameras or motion sensors, and with little to no cover/concealment.

With a low crawl, movement is tiring and slow. Lots of gear will make things worse, so try not to overpack and crawl only when necessary. In environments with wide-open spaces and little cover, try not to pack your tactical vest with too much gear on your belly and back. You want to maintain a slim profile when dropping to a low crawl.

High Crawl

A high crawl looks similar to a low crawl, but with the body and rear end in a higher plank position. Where low crawling looks more like dragging your belly, knees and elbows are used in high crawling to move progress forward. High crawling enables a quicker pace and tends to be a little less tiring. However, it increases one's visibility and should only be used where some cover and concealment are present. For example, terrain with high grass, small shrubs, and some tree cover.

If my team is going to resort to crawling, high crawling is usually the technique of choice. We make it a point to choose routes that involve the most cover and concealment so there isn't a need to hit the deck.

One important note to be made about crawling is that it will get clothes dirty. I recall a mission where two of my team were to change from their tactical clothes at the perimeter and into "staff clothes" so they could blend in with the late-night staff after they scaled the fence. It turned out we all had to hit the deck and high crawl for a short time in our tactical clothes before reaching the fence. Our clothes were dirty and sweaty. Thankfully, a change of clothes was already in the plan.

Rush

Unlike crawling, the rush is the faster method to get from one point to another. This, however, should not look like a full-on upright run. In appearance, it looks much like a crouched brisk walk. Each rush should last about three to five seconds. Quick rushes make it difficult for potential bystanders to fully track movement. Movement may unintentionally catch their eye but should not hold it long enough to be convinced of movement. Breaking up rushes with at least five seconds between rushes will help avoid unwanted attention.

Each rush must have a designated destination. In other words, an operator should not drop to the ground simply because five seconds have passed. A drop destination providing some level of cover or concealment should be sought out in conjunction. Forethought as to the next drop position is mandatory.

To re-iterate, even when the coast seems clear, I highly advise against rushing for more than five seconds. While I've mentioned being aware of one's silhouette and 360° profile during movement, it is quite a difficult skill to master. I've fallen victim to rushing the perimeter of a building for longer than I should have. It's tempting to just want to go for it. Instead, adopt the five-second rule, drop, and take the time to cycle through an OODA loop.

Signaling

The use of hand signals is commonly used by the military and certain branches of law enforcement. They are useful to red teams for communicating movements during periods of an engagement where silence must be maintained in order to conceal the team's presence. Signaling is also a good backup plan when radio communications fail. While they may get little use in the field, they should be learned and practiced.

When signaling is in use, the red team operator who is first in line typically becomes the signal giver. Of course, there is no hard and fast rule, so this role can be given to anyone in the team. What's most important is that the signal giver takes the point position and the rest of the team is able to clearly view the signals.

There are quite a number of signals to borrow from. However, the following pages illustrate those that I find most relevant to physical red team operations.

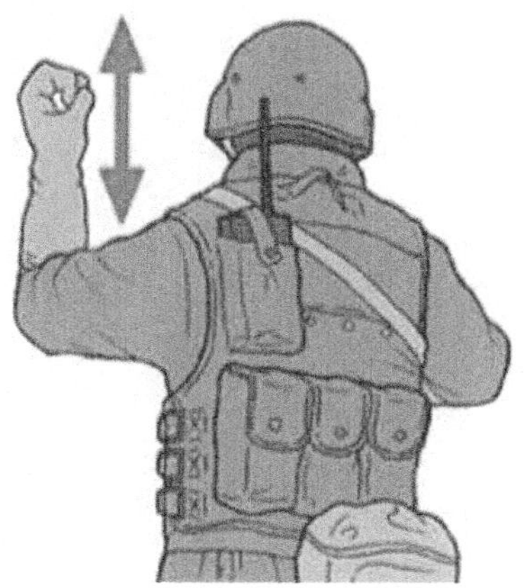

Figure 39. Hurry Up

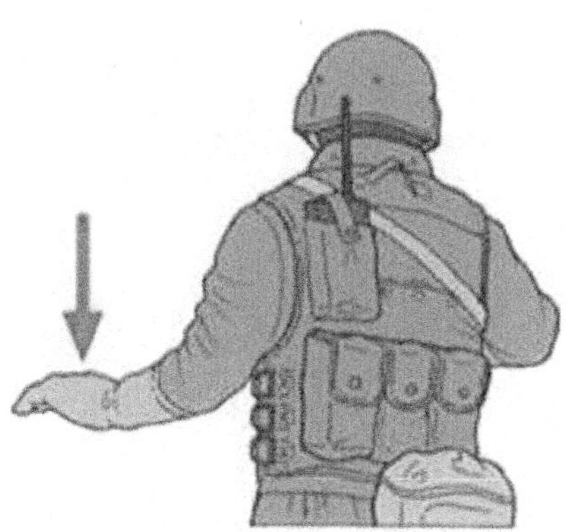

Figure 40. Crouch, Go Prone, or Crawl

Figure 41. Freeze

Figure 42. File Formation

Figure 43. Vee Formation

Figure 44. Go to Rally Point

Figure 45. I Understand

Figure 46. I Do Not Understand

Maneuvering an operation with precision and efficiency requires knowledge about the environment, settlement, and how to leverage surrounding cover/concealment to one's advantage. Quick decision making, clear communication, and careful movement are all part of what makes this leg of the mission important.

For hands-on Physical Red Team Training, please visit:

https://www.redteamsecuritytraining.com/physical-red-teaming-book

[CHAPTER 9]

OFFENSIVE STRIKE

Both teams, having successfully made it to the exterior wall of the building, radio the red team leader with a SITREP. Alpha team states they have reached the loading dock door. Bravo team announces they have reached their position at an employee entrance to the rear of the building. The suspected vulnerabilities: a loading dock door that is believed to be poorly hung and an employee side entrance the cleaning crew usually leaves unlocked during their night shift.

The red team leader replies over the radio, "copy that." A member from alpha team takes a knee and reaches into his tactical bag for a Shove-it tool. Feeling the sweat under his clothes, another shot of nervousness blasts through his body. He can see the silvery reflection of the metal lock in the door frame and pushes the Shove-it in. Also kneeling and facing the opposite direction is the second member of the alpha team keeping watch. Meanwhile, bravo team hides behind bushes just outside the employee entrance door. They are listening for activity and, just then, a cleaning crew worker walks out for a smoke break. Bravo team hits the deck, and they wait. The smoker finishes and re-enters the building--by badging in. Badged entry? On this door? Is this a new security control?

At the other end of the building, alpha team works the loading door lock. Suddenly, the handle gives way and the door opens slowly. They were right about the door. Before them, pitch black darkness, a cavernous sound, a strong smell of diesel, and absolutely no idea what lies just in front of them. Welcome to Offensive Strike.

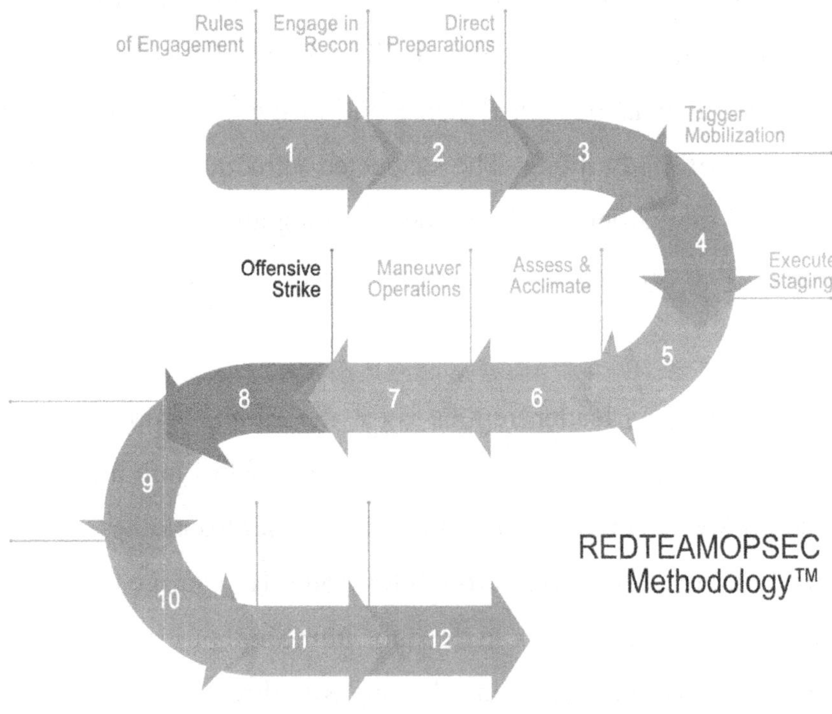

Figure 47. Offensive Strike Phase

The Offensive Strike phase is always chock full of surprises. Next to the Evacuate, Evade, & Cover phase, it's one of the most exhilarating. This stage is where the hypothetical becomes reality, where suspected vulnerabilities are either found to be exploitable or not.

Since this chapter is predominately about exploitation, I will concentrate on the physical security controls that I encounter the most and possible ways to defeat them. Note, this is far from an all-inclusive list.

Ground Sensors

Figure 48. Unattended Ground Sensing System

Ground sensing systems, sometimes called Unattended Ground Sensors (UGS), use technology such as seismic, acoustic, and magnetic sensors to automatically detect the presence of people or vehicles. When sensors pick up activity, they usually transmit alarm data to a control hub via radio frequency (RF). Control hubs then transmit to a central control center, often a nearby security operations center (SOC), for incident response teams to manage. Other UGS systems exist with more advanced technology, however this type of system is what we commonly come across.

Ground sensing systems use hardware and cable sensors that are usually buried beneath the surface. Burying the system hardware helps prevent unwanted tampering or disarming and aids in concealment as well. UGS systems are usually placed in key areas of a facility's external perimeter and are often meant to stay there for long periods of time.

Here are a few characteristics of a UGS implementation:

- Difficult to visually detect; relevant in low-traffic areas

- Usually intended to "cover" a lot of ground area

- Usually placed a few feet outside a security fence

- Often used as a replacement for guards, guard tours, 24/7 eyes-on cameras, and other motion sensors

- Detection rates can vary greatly according to buried cable depth, cable type, vendor, and implementation tuning

- Extremely prone to false-positives

Figure 49. Cabling Hardware for UGS

Identification

The most common type of systems we encounter are seismic systems that recognize vibrations in the ground. That said, it is very difficult to visually identify UGS systems. Unless red teamers manage to uncover open source intel or first-hand intel about the presence of UGS at a target, the only real way of knowing is by bait testing for it. You can bait test by walking near or on the suspected area with a plausible pretext. During one engagement, my team found a nearby Humane Society and volunteered to walk dogs. They walked a dog very close the suspected area. The dog-walker pretext offered a plausible alibi, if stopped, and enabled the team to have a much closer view of the facility. Later in that engagement, we ran another bait test by fast-walking over the suspected area and back to tree cover. I did this test repeatedly until we were satisfied. Be aware, this kind of bait test involves more risk and should only be performed if there is a pre-established pretext and cover/concealment is available.

UGS systems are a great physical security control for several reasons. They are difficult to detect and not easy to hack. You may never know one is in place until it's too late. But don't let that notion fester in the pit of your stomach. UGS systems are not the golden security control that people make them out to be. If they are poorly implemented, not continually tuned to the natural movement of the environment, or not maintained, they are less effective. And their detection success rates vary by vendor solution. UGS systems have one enormous flaw. They rely on *people* to make them effective. That's right. They only work if people respond the right way every time.

Ask any experienced cyber security person their feelings about working with their company's network intrusion detection system (NIDS), and I would be surprised if you're not met with sighs and eye rolls. UGS systems are no different in principal to NIDS. They require constant care and feeding, and they regularly spew annoying false alarms in the middle of the night. Oftentimes, there are so many false alarms that security people simply begin to ignore them.

BINGO!

Bypass & Defeat

Taking a shovel and pick to a buried UGS system is not the right approach. The most effective tactic toward defeating UGS systems is by way of its responders. My team does this by creating several alarms to fool the responders into thinking there is a glitch in the system, which they later begin to ignore. I describe this false alarm tactic in the first chapter of this book taken from my interview with a reporter with The Houston Chronicle: https://www.houstonchronicle.com/business/article/Put-to-the-test-cybersecurity-experts-easily-10989830.php.

It makes no difference if you pound the ground with a rubber mallet or a rubber horsehead. The key to the false alarm tactic is persistence and avoidance. Red team operators must be close enough to be detected and remain unseen when the first responders arrive. The false alarms should continue while the responders are onsite and well after. Personally, I've continued this tactic for nearly two hours straight. Persistence makes the tactic more convincing and increases the likelihood responders will ignore the alarms, providing red teamers an open window for exploitation.

Fencing

Figure 50. Anti-climb Fence

Anti-climb fences are among the most common type of fences, aside from the typical chain-link fence. With substantial space between the links for fingers and toes, chain-link fences are easily climbable. Anti-climb fences, however, have a narrow wire mesh that makes climbing with fingers and toes very difficult. Almost all anti-climb fences have this narrow mesh design, while some also utilize barbed wire or spikes on top.

Identification

Here are a few characteristics of anti-climb fencing:

- Rectangular narrow wire mesh

- Thick vertical iron bar design, often with angled spikes on top

- Angled and irregular patterned wire mesh design

- Chain-link with hard plastic material woven in

- Razor wire, barbed wire, or angled spikes on top

- 8 feet to 18 feet high

Essentially, fences are designed to slow down an attacker's advancement and potentially inflict fear of injury. Anti-climb fences look intimidating because they're supposed to look intimidating. To physical red teamers though, they are merely one of the nominal challenges they are likely to face during a mission.

With the right tactics and tools, just about any security fence can be exploited. Let's examine a few simple ways to bypass anti-climb fences.

Bypass & Defeat

Let's start with the obvious and definitely the most used by my team: Ladders. Operator #1 places a ladder against the fence and climbs up. Operator #2 hands Operator #1 a second ladder which is placed on the opposite side of the fence. You can probably guess what happens next.

Figure 51. Defeat Security Fencing

But what about the scary barbed wire, razor wire, and spikes? Carpet remnants or thick wool blankets placed over the top will prevent injury. My team uses standard-issue U.S. Army wool blankets, but any pliable yet highly thick fabric will do.

Factors to consider when using this tactic:

- Exercise with extreme caution

- Operators <u>must</u> be physically agile

- Have around 4 ft. x 4 ft. of durable fabric to prevent injury

- Wear ripstop clothing, durable boots, and gloves

- Rehearse this tactic before using in the field

This bypass tactic can be dangerous and should only be carried out with the proper training and safety measures.

Another less popular tactic is to utilize specialized climbing gear. Believe it or not, ninja hand and foot claws can make climbing an anti-climb fence possible. I say this with caution though. The hand claws are made of durable steel, as are the foot claws. However, they can do quite a painful number on unprotected hands. For this to work properly, additional padding absolutely must be added so that your hands do not feel like they are about to separate from your arms.

The ninja hand and foot claws are very sharp, and odds of injury are high. This should only be carried out as a last resort and by operators in excellent physical condition.

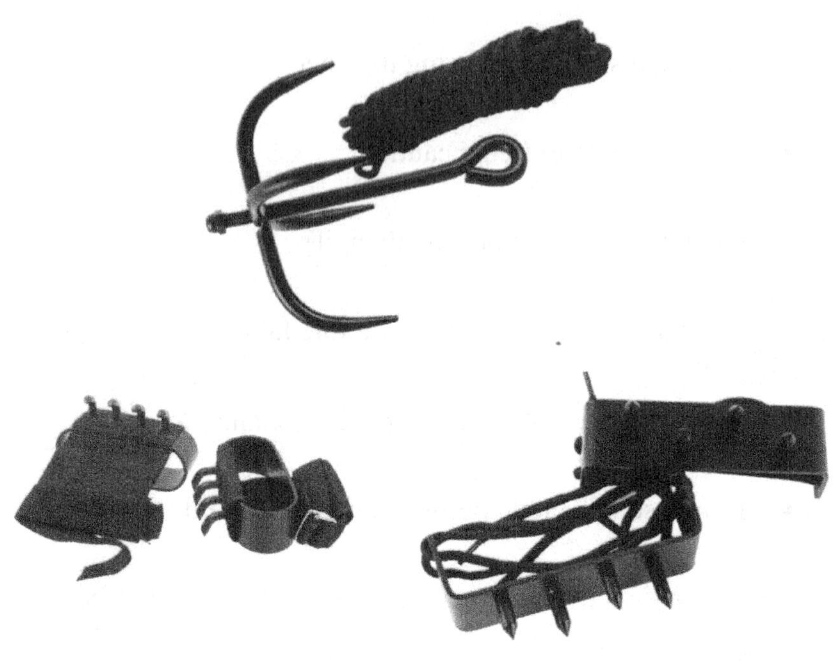

Figure 52. Climbing Gear

Motion Sensors

An electronic motion detector contains an optical, microwave, or acoustic sensor. However, a passive sensor recognizes a signature only from the moving object via emission or reflection. For example, it can be emitted by the object or by some ambient emitter, such as the sun or a radio station of sufficient strength. Changes in the optical, microwave, or acoustic field in the device's proximity are interpreted by the electronics and can trigger an alarm or series of actions.

Motion detectors have found wide use in domestic and commercial applications. A motion detector may be used to alert a homeowner or security service when it detects the motion of a possible intruder. Such a detector may also trigger a security camera to record the possible intrusion.

Microwave sensors detect motion through the principle of Doppler radar and are similar to a radar speed gun. A continuous wave of microwave radiation is emitted, and phase shifts in the reflected microwaves due to motion of an object toward (or away from) the receiver result in a heterodyne signal (two signals combined into one) at a low audio frequency.

In an ultrasonic sensor, a transducer emits an ultrasonic wave (sound at a frequency higher than a human ear can hear) and receives reflections from nearby objects. Exactly as in Doppler radar, heterodyne detection of the received field indicates motion. The detected doppler shift is also at low audio frequencies (for walking speeds) since the ultrasonic wavelength of around a centimeter is similar to the wavelengths used in microwave motion detectors. One potential drawback of ultrasonic sensors is that the sensor can be sensitive to motion in areas where coverage is undesired, for instance, due to reflections of sound waves around corners. Such extended coverage may be desirable for lighting control, where the goal is detection of any occupancy in an area. But for opening an automatic door, for example, a sensor selective to traffic in the path toward the door is superior.

Passive infrared (PIR) sensors are the most common to us and what we will concentrate on here. PIR sensors are sensitive to a person's skin temperature through emitted black-body radiation at mid-infrared wavelengths, in contrast to background objects at room temperature. No energy is emitted from the sensor, thus the name passive infrared. This distinguishes it from the electric eye for instance, in which the crossing of a person or vehicle interrupts a visible or infrared beam.

Identification

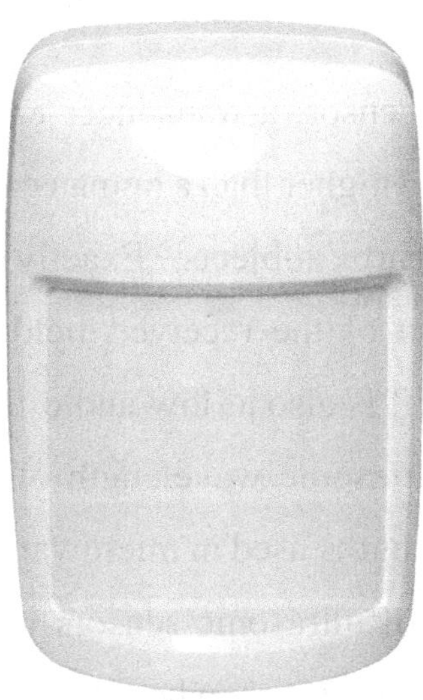

Figure 53. Motion Detector

Though motion detectors come in all shapes and sizes, they tend to share a common form factor. The "eye," or actual sensor, is identifiable by a spherical shape or behind a window, as shown in the example. Low-cost detectors have a range up to 15 feet while others offer much longer ranges.

Again, from my experience, most of the motion detectors I encounter look like Figure 53, use PIR sensing technology, and have a range of about 15 to 25 feet.

Bypass & Defeat

Figure 54. Thermal Radiant Film

My team uses a homemade infrared (IR) shield using thermal radiant film stretched over a wooden square frame large enough to hide behind.

Figure 55. My team with homemade IR shield (Credit: Paul Szoldra/Tech Insider)

Of utmost importance, of course, is to build handles on the hidden side so as not to leak body heat from hands and fingers while holding it up. The IR shield tactic has worked successfully on many occasions and is my team's go-to solution.

On a side note, I have seen a video on YouTube of someone successfully bypassing an IR motion detector by holding up a white sheet instead of thermal radiant film. Even though it worked in the video demonstration, I recommend using a more robust solution with heat resistant film instead.

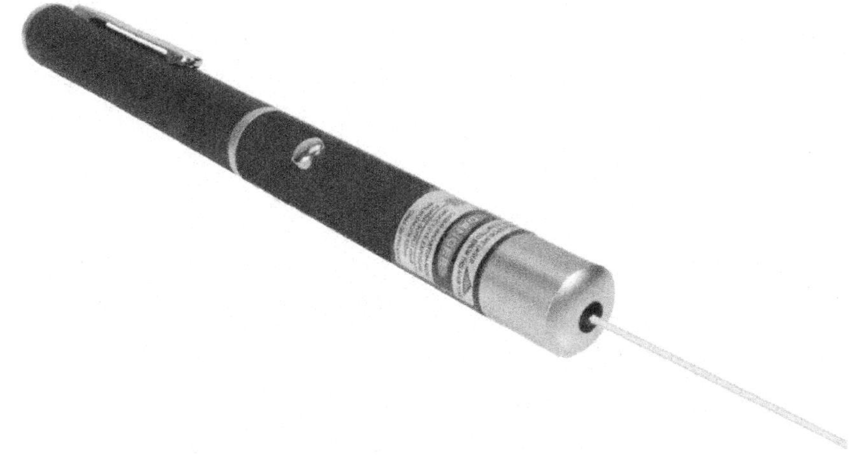

Figure 56. High Powered Laser Pointer

Strong laser pointers can be used to essentially blind PIR devices in order to prevent them from triggering. The tactic involves simply directing the laser beam at the center of the PIR device's eye. Carrying out this tactic requires a very steady hand, however. For long distances, a tripod should be used to steady the beam. This tactic is really only useful for blinding one PIR at a time. For these reasons, this tactic is not an option we go with very often.

As a final resort, some PIR motion detectors can be thwarted by simply moving very, very slowly through the detection area. Since most detectors are mounted high, crawling instead of walking can help increase exploitability. But again, an operator must move very slowly to be effective and the mission may not allow for that kind of time.

Alarms

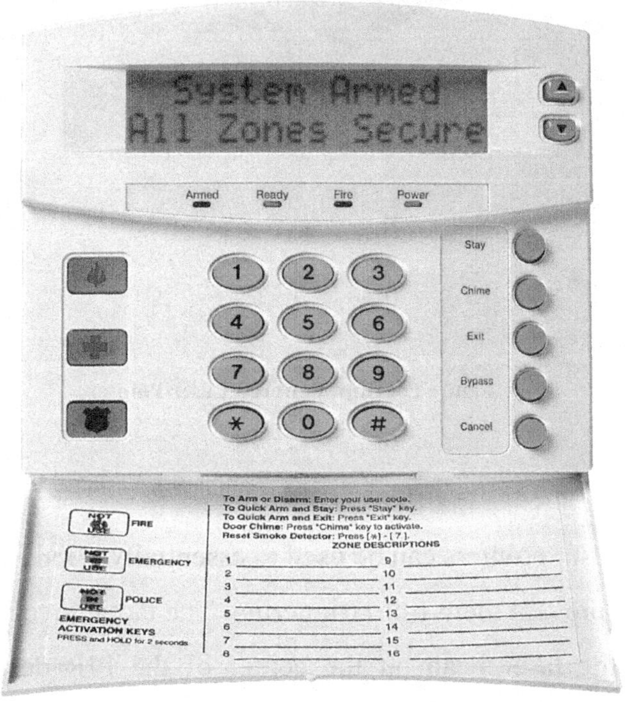

Figure 57. Typical Alarm Control Center

Many of today's commercial alarm systems rely on the same type of underlying technology used to protect residences as well. Albeit, commercial systems typically include PIN pads, RFID readers, request-to-exit (RTE) sensors, and much more. What is similar about these systems is the communication medium used to relay data from the sensors to the primary controller and from the controller to an authoritative alarm response center (ADT, law enforcement). Wireless technology, such as Wi-Fi, 4G, 3G, GSM, 433/315/868 MHz RF, has replaced many of the old hardwired systems.

Identification

An operator will see an alarm sensor or ten (see Figure 53) before ever seeing the alarm control panel and its brand/model. Most commercial control panels, not to be confused with keypads, are installed out of sight in a utility closet or server room. Thus, alarm system identification isn't always feasible.

In reality, the brand of alarm system is not as important as the technology it uses to detect and communicate. What we are most interested in in this section is the technology it uses to communicate to other sensors/sirens and its alarm response center.

Bypass & Defeat

As I mentioned earlier, most current alarm systems use RF and/or Wi-Fi to communicate locally and a variation of GSM, Wi-Fi, or 4G to alert the authoritative alarm center externally. Vendors will co-mingle and intermix all sorts of technologies together in various models of alarm solutions for their customers. So even if you know the brand of the alarm solution, you may not know the exact communication medium it uses. What is a red teamer to do?

Figure 58. Signal Blocker

Signal blockers are used by attackers to degrade and sometimes completely block alarm signals. The signal blocker pictured here has twelve antennas that can isolate and block GSM, 4G, LTE, Bluetooth, Wi-Fi, 433/315/868 MHz, CDM, 3G, LOJACK, 5G Wi-Fi, and GPS separately. Many alarm systems and their sensors operate in this very space. Thus, signal blockers are a real threat to alarm systems and prove to be one of the most effective ways in bypassing them.

Signal blockers are illegal in the United States, according to the FCC, and I do not advocate their use where prohibited.

Doors & Locks

Doors and locks make up the majority of the physical security controls my team confronts. It would take a volume of books to cover the variation in locks, doors, levers, knobs, and their respective vulnerabilities. But in the ongoing spirit of this chapter, I will address the doors and locks my team meets every day.

Identification

When it comes to doors, we do not immediately resort to lock picking. Lock picking takes time, it is noisy, it can give away your position, and it looks nothing like in the movies. So just like it's done on the cyber side, we first look for vulnerabilities. What kind of door is it? Is it old or new? Where are the hinges? What kind of handle does it have? How is it hung? By visually scanning for vulnerabilities or lack thereof, we determine which exploitation route is optimal--to pick or not to pick. Generally speaking, we usually try to bypass it instead.

Figure 59. Levered Handle

The levered handle door is very common in businesses from offices to warehouses. Reason being, its physical configuration is governed by the Americans with Disabilities Act in the U.S. Specifically, the ADA has requirements for the amount of tension applied to activate the door lever, to its height from the floor to the amount of pressure needed to open the door.

Figure 60. Set of Crash Bar Doors

We have all seen these types of doors, particularly in hospitals, shopping malls and large enterprise complexes. They are sometimes referred to as panic bars, push bars, and exit bars. They too, have specifications governed by the ADA ensuring they can be used by all.

Crash bars are more commonly found in the lobbies of buildings to allow for a mass exodus of people in cases of emergency. They will be scattered through the internals of a building, where maintenance workers can open them while pushing big trash bins or crash carts. They are also very popular as designated emergency exit doors.

Figure 61. Commercial French Door

The commercial French door with crash bar activator is a very popular configuration in most businesses. The center gap between the doors is what's most interesting to us red teamers, but more on that later.

Figure 62. Standard Door Knob and Lock

The other type of door handle is the standard knob. In a business setting, you may not run across many standard door knobs because they do not meet ADA compliance. Standard knobs are typically found on utility closets, network closets, storage rooms, special entrances, service doors, etc.

Figure 63. RFID Controlled Door

Many companies these days use Radio Frequency Identification (RFID) technology to control access into their facilities. RFID uses electromagnetic fields to automatically identify and track tags (RFID card as shown in Figure 63) attached to objects. The RFID card contains electronically stored information, much like a unique serial number. In an RFID access control system, this unique serial number is linked to an individual or group of individuals. From there, access into areas of a facility can be managed electronically for that individual for all doors that are RFID-enabled. As seen in Figure 63, HID is the dominant RFID access control solution provider in this space.

Lever Handles

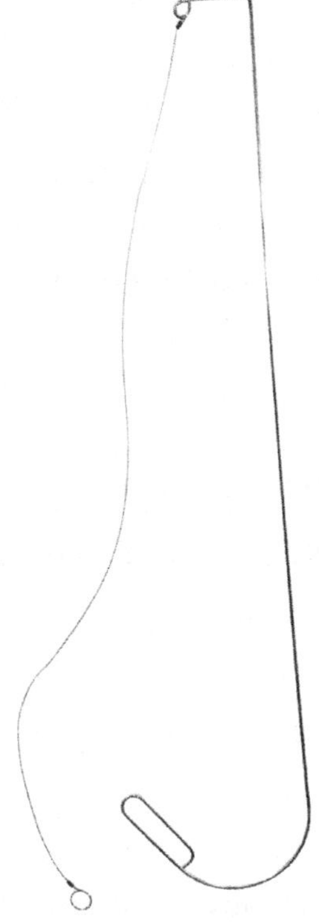

Figure 64. Under The Door Tool

Doors with levers are susceptible to bypass using a tool appropriately named the Under The Door Tool (UTDT). My team has used this on countless engagements with great success. This tool is a must-have!

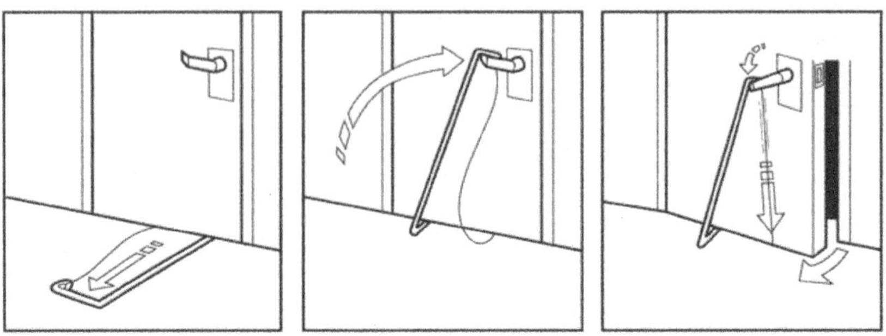

Figure 65. Under The Door Tool Instructions

It is an odd-looking piece of equipment that not only takes a little getting used to, but it takes a minute to wrap your head around how it works. To help, I recommend a quick search on YouTube for a visual demonstration.

Crash Bar

Just like the standard lever handle, a tool exists aimed at exploiting the flaw inherent in crash bars. The **double door bypass tool** exploits the gap between French doors equipped with crash bars on the inside of the doors. See Figure 60. First, the bypass tool is inserted in the gap between the two doors. Most doors will have rubber weather stripping or brush material in between. Once most of the tool is through the gap, the operator turns the tool 90 degrees and lines the tool up with the crash bar on the opposite side. Then, she pulls the tool inward, thereby depressing the crash bar and opening the door.

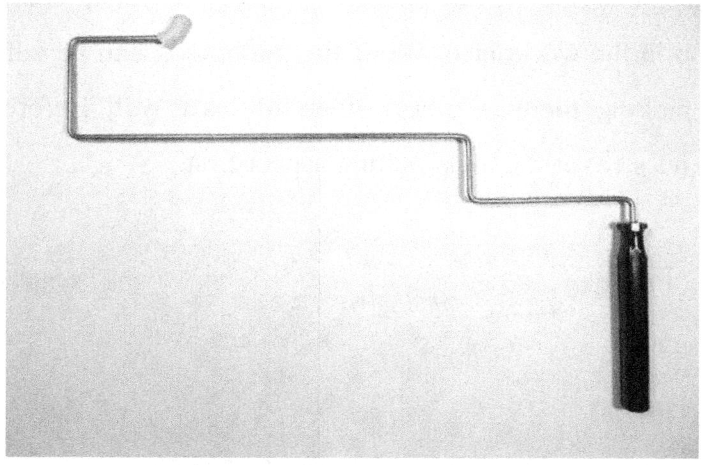

Figure 66. Double Door Crash Bar Tool (Photo credit: Rift Recon)

To better understand how this tool works, I recommend a quick search on YouTube for a visual demonstration.

On a side note, crash bar tools can be made on the cheap with moderately gauged wire and a vice. Otherwise, visit your local Home Depot.

Door Knob (Lock)

Door knobs like the one pictured earlier in this chapter are pretty common in the workplace. While the knob itself can be vulnerable to lock picking, there are other vulnerabilities as well. Before we get to that, let's cover the lock picking aspect first.

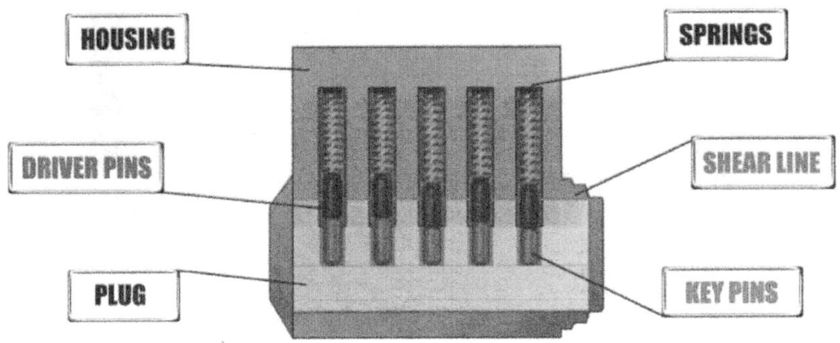

Figure 67. Pin & Tumbler Lock Anatomy (Credit: Art of Lock Picking.com)

There are entire books devoted toward mastering lock picking. This book will barely scratch the surface. Instead, I hope to introduce some fundamentals to spur further learning on the subject.

In a pin and tumbler lock, the most common lock my team faces, the springs maintain a downward tension on the driver and key pins. This ensures the driver pins are always blocking the shear line, which prevents the lock from opening. See Figure 67. When the right key is inserted into the keyhole, the key pushes the spring-loaded key pins higher up in the housing. The correct key's peaks and valleys (bitting cuts) match with the irregularly-sized key pins to lift the driver pins up and align perfectly to form a straight horizontal shear line. Again, with the correct key, a straight horizontal shear line is created, allowing the key to open the lock.

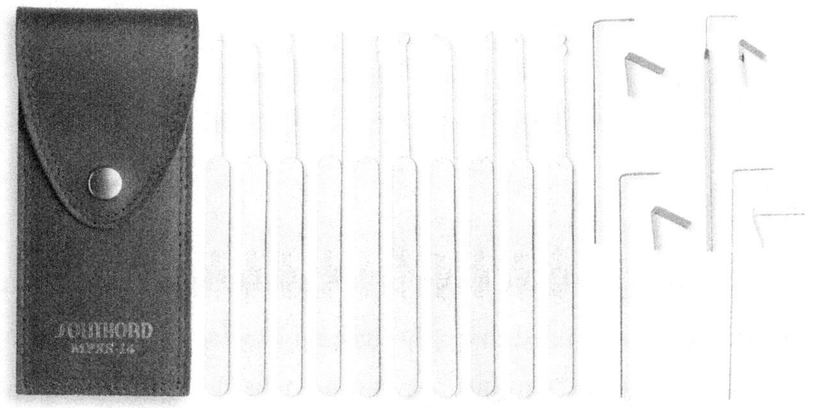

Figure 68. Sample Lock Pick Set

Picking a lock, however, is made possible by exploiting manufacturing defects in the machining of the lock so that the pins can be agitated and torqued enough with a pick and tension tool to make the driver pins sit askew inside the housing. Clearly, this is not the manufacturer's intent, and some manufacturers go through extreme lengths to prevent picking. Additional pins, security pins, and different shaped pins are a few examples of mitigating controls manufacturers use.

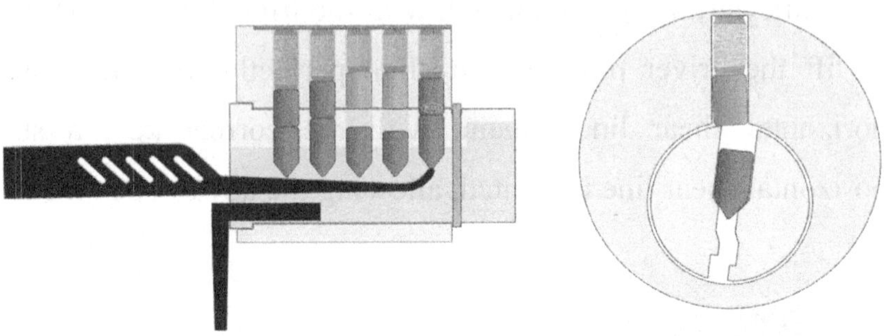

Figure 69. Pick and Tension Tool Causing Pin to Bind

Figure 69 shows how a lock pick is used to agitate a key pin, and the tension tool is used to apply light clockwise torque, causing the driver pin to bind. I highly recommend bookmarking a YouTube channel by 'BosnianBill' for a more in-depth study from basics to advanced lock picking.

Lock picking resources:

- BosnianBill: https://www.youtube.com/user/bosnianbill

- TOOOL: http://toool.us/resources.html

- Hacker Warehouse: https://hackerwarehouse.com

- Sparrows Lock Picks: https://www.sparrowslockpicks.com

Figure 70. Shove-it Tool

Stepping away from lock picking, there are many doors that are vulnerable to shimming. Have you have ever seen someone crack open a locked door with a credit card in a movie? Technically, that's shimming. Doors that have small to large gaps between the door and the frame could be vulnerable. You stand a good chance of being able to shim a door if you can see the slight metallic reflection of the locking mechanism through the door and frame. See Figure 71.

Figure 71. Shim Exploitable Door

There are several tools designed specifically for shimming, ranging from plastic to metal to wires. In reality, they all do pretty much the same thing. My tool of choice is the Shove-it Tool (see Figure 70).

The Shove-it Tool is a simple lock bypass tool that works on many types of locks. The shape of the tip allows for pushing, pulling, or sliding latches. A red team operator would simply slide the device into the gap between the door and the frame to activate the latch and open the door (see Figure 72).

Figure 72. How a Shove-it Tool Bypasses a Latch

Before we assume that all doors are vulnerable, consider that some lock manufacturers have put controls in place to deter shimming. Notice the half-moon shaped metal piece to the left of the latch. That is sometimes referred to as a tamper pin. When the door is installed properly, only the latch should go into the hole in the metal plate on the frame (keeper) and the tamper pin should be depressed. When the latch sits inside the keeper and the tamper pin is depressed, shimming becomes more difficult. Yet we see most latches with tamper pins installed to allow the tamper pin to go inside the keeper, making shimming much easier.

Metal shields designed to cover a door latch (strike plate cover) to deter from shimming a door are popular with installers. A simple but effective approach is to use a longer Shove-it Tool.

Figure 73. Large Strike Plate Cover Over Large Door Gap

RFID

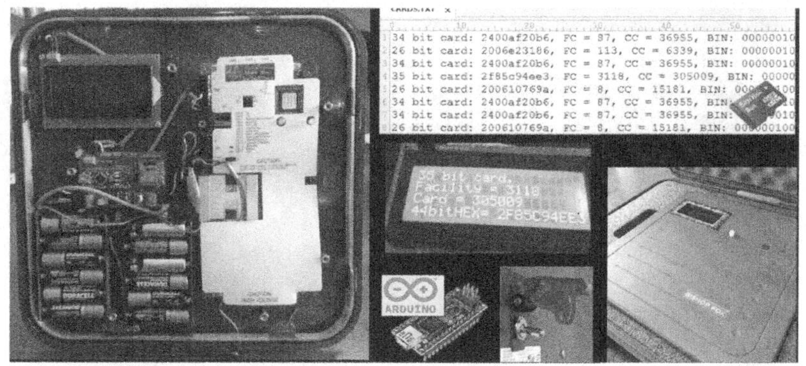

Figure 74. Tastic RFID Thief by Bishop Fox (Photo credit: Bishop Fox)

As I mentioned earlier, organizations make heavy use of RFID readers on doors and issue RFID cards to employees to electronically manage access in and out of their facilities. These systems make access control very efficient and secure for businesses when implemented properly. However, many of today's organizations are unaware of the risks of using an insecure RFID implementation. Tools like the Tastic RFID Thief in Figure 74 make stealing RFID access from employee badges trivial. My team has built several readers like this one using parts and schematics available widely on the Internet.

An RFID reader tool is a must-have in every red teamer's kit. Figure 74 shows a modified 12x12 inch HID RFID reader that has undergone massive repurposing for RFID stealing.

Figure 75. My team with RFID in the field (Credit: Paul Szoldra/Tech Insider)

Figure 75 shows my team member using an RFID reader hidden inside a laptop bag. The red teamer scheduled a meeting, under false pretenses, with an employee known to have an RFID badge with elevated building access privileges. The red teamer got close enough to the target's badge to later make a duplicate copy which was used to gain access into the building later that night.

Stealing and cloning RFID employee badges is a real and rampant risk. Nearly all of my team's physical red team engagements have involved use of our RFID tools to some extent. Operation of the RFID reader is fairly straightforward if you are somewhat technically savvy. Where the real rubber hits the road is how creatively an operator can use and covertly disguise one to achieve their goal.

Acquiring an RFID reader like the one depicted here is not always easy. So I've provided a few resources below to help those new to the technology get started.

RFID reader and cloner resources:

- https://www.bishopfox.com/resources/tools/rfid-hacking/attack-tools/

- https://www.youtube.com/watch?v=W22juSqhJSA

- https://lab401.com/collections/hardware/products/rfid-pentester-pack

Please visit the links below to learn more about how my team used an RFID reader during a real physical red team operation.

https://www.businessinsider.com/red-team-security-hacking-power-company-2016-4

https://www.businessinsider.com/clone-rfid-security-badge-2016-5

[CHAPTER 10]

PENETRATE & CONTROL

The alpha team gets ready to make entry through the now shimmed loading dock door. Pulling an infrared borescope from his tactical bag, one operator adjusts the camera fixed to the stiff gooseneck to covertly peer inside and all around the door. Satisfied no immediate risks are present, both operators reach for their night vision monocles, crouch down, and prepare to cross the threshold.

Meanwhile, the door that bravo team planned to compromise has been recently outfitted with an RFID reader. Recognizing the door is ADA regulated, they devise a risky but alternative strategy. One operator, who had planned to change into street clothes after entering, changes now and stands on the sidewalk near the entrance. The other operator moves into a crouched position on the opposite side of the door. Several minutes pass that seem like a lifetime. Then, the loud and unexpected noise of the latch opening nearly scares the team, and a crew worker begins to exit. The street-clothed operator immediately hollers to the crew worker asking to bum a cigarette as the other operator quietly ducks around the door and inside. The ADA regulated door speed held the door open with just enough time to sneak through.

With the help of night vision, the alpha team winds its way through a maze of pallets stacked eight feet high before seeing a set of double doors. It's dark, and nobody is supposed to be in here. They constantly scan for motion detectors and cameras as they make their way to the doors. It feels like walking through a minefield. Finally, they reach the doors. Posted on the wall is an aerial map of the building's emergency exits. They catch their bearings and advance. This is Penetrate & Control.

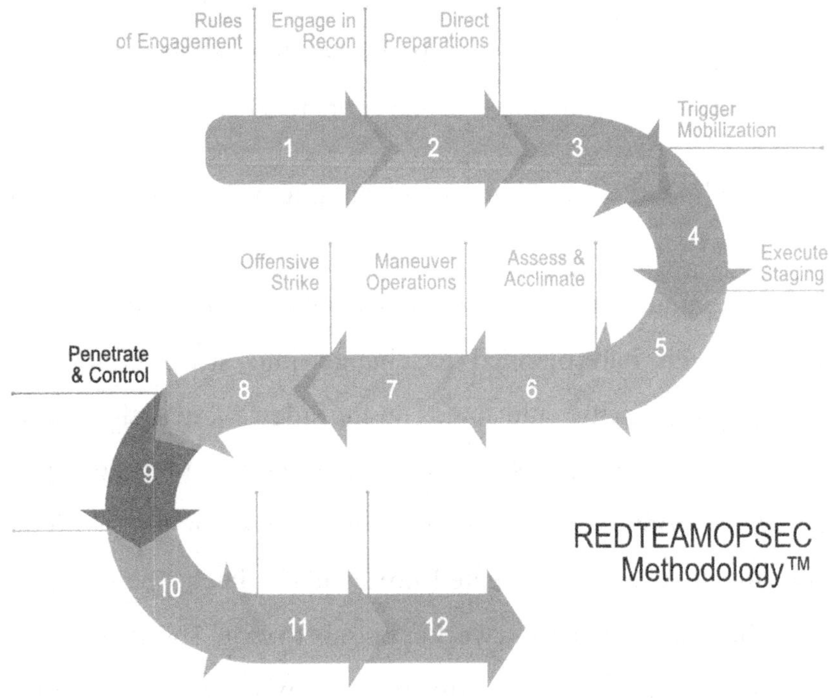

Figure 76. Penetrate & Control Phase

There's nothing like breaking into a building in the middle of the night not knowing who or what is around the next corner. Even though these are lawful engagements, the feeling isn't any less nerve-wracking and adrenaline-charged. How do I go unnoticed? Are there people in here right now? Where am I and which way should I go? What is around that corner? What if someone sees me? As physical red teamers, we've all asked ourselves the same questions. Days upon days of planning has led up to this and any little mistake could be devastating to the operation.

This phase offers much needed guidance on how to penetrate and control access to a facility.

Character Change

Changing clothes and character from a red team operator to an office cleaner or employee persona, for example, is not uncommon once building penetration has been reached. Once inside the building, switching from black tactical clothes to a business casual employee costume can be a great pretext to fall back on if spotted. In fact, some operations may require a character change.

Here are some situations that may warrant such a change:

- If the facility will be occupied during Penetrate & Control

- If the potential for occupancy is not known

- If the facility's internal layout is large and not known

- The operation will take place overtly or during the daytime

Precisely which character to change to is entirely dependent upon plausibility and which persona (office cleaner, employee) is likely to occupy the facility at the given time.

Figure 77. Cleaning Smock

A character change almost always involves changing clothes and developing a pretext. The complexity of both depends on the level of security awareness of its staff, who may stop and question the team. For offices, the most common persona is the commercial cleaning worker. It can be pulled off with relatively simple clothing and props. Jeans, a smock, and some latex gloves. Add a cleaning tray to store your pick set, flashlight and spray bottle or two.

Figure 78. Cleaning tray as a prop

I recommend using a clothing change and pretext as a backup plan to nearly every operation.

Establish Your Position

Operational orders (OPORD) almost always require red teamers to reach a destination, like a server room, and perform a number of tasks once inside a facility. But how do you get to that destination when you don't know where it is? This is often the case with me and my team. Sometimes we catch a break and find the building layout ahead of time through reconnaissance. Sometimes the building's external layout makes it evident. But generally speaking, we never know the facility's floorplan until we are physically in the midst of it.

Cardinal Direction

First and foremost, every red teamer must understand their four cardinal directions and how to find their position with a compass. The four cardinal directions, or cardinal points, are the directions north, east, south, and west, commonly denoted by their initials N, E, S, and W. Points between the cardinal directions form the points of the compass.

Most smartphones and smartwatches can do this pretty easily with the help of GPS. However, I always caution the use of these devices because they usually need to be activated and emit a bright light, which may give away your position in a dark area. Instead, I recommend using a wrist compass with night glow. I highly recommend this for those of us who are abnormally directionally-challenged, like my wife. Bless her heart.

Figure 79. Glow in the dark wrist compass

A wrist compass keeps an operator's hands free, doesn't require activation, and doesn't emit any light in the process. Using a compass to orient oneself and obtain their bearings using a memorized aerial photo of the building is very effective. If necessary, a small aerial printout could be carried by the operator if the building happens to be a sprawling complex.

Emergency Maps

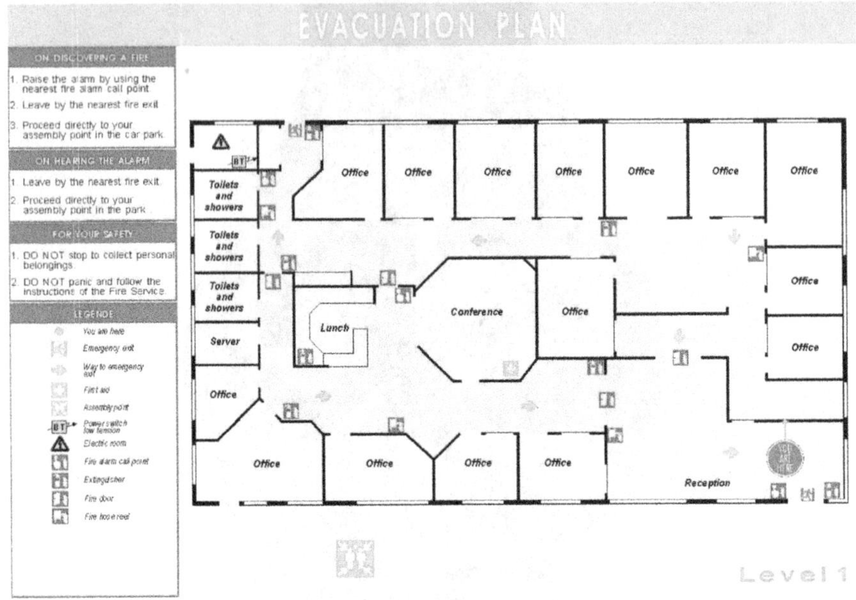

Figure 80. Emergency evacuation map

Safety administrations, like OSHA in the U.S., mandate certain safety requirements for businesses. For example, OSHA requires organizations to develop an emergency action plan for the goal of protecting lives and property during an emergency; an evacuation policy that provides posted signs and placards concerning emergency exits, fire extinguishers, first aid kits, and so on. While floor maps are not specifically identified, many businesses choose to convey this information using a posted floor map. I have some good news and some bad news. First, the bad news. Not all businesses are required to convey emergency information using building maps, and when they do, the amount of detail can vary greatly. Now for the good news. It's much easier to convey information visually using a building map, and we see that most businesses do.

Feel free to jump for joy when you spot one of these little gems! Generally, no matter how little information it may provide, it is usually better than nothing. Use what information you are able to glean from posted signs to support establishing your position and direct you where to go

Movement

Take another lesson from the previous chapter, Maneuver Operations. Movement through a facility, under covert conditions, should be done using the rushing technique, just as movement should be done outside a facility.

Rushing is carried out by slightly crouching at the waist, bending at the knee while keeping the head facing forward (see Figure 81). This makes an operator's profile smaller than walking upright, yet it enables quickly dashing from one position to another. I have found that rushing enables me to hide the sound of my footsteps a little better. But again, rushing should only be used during covert movement and in areas where there is little to no chance of the area being occupied. It would be hard to smooth talk your way out of being seen creeping around suspiciously like that.

Hazards

Figure 81. Avoiding hazards while rushing

There are all sorts of hazards that could potentially give away the position of a red teamer and make their task harder. If I had to list the most critical hazard when it comes to penetration and control, I would say windows. In the throes of an infiltration, it is difficult to be situationally aware of 360° around your body, and it's easy to walk right past a window that could give you away. Unless an office lobby is the objective, they should be avoided for these reasons. But there is more than just one hazard to be aware of.

Here is a list of the most common hazards:

- Windows

- Doors, corners, and stairs

- Cameras and motion detectors

- Lighting (internal lights or poor operator light discipline)

Let's briefly talk about lighting. Earlier in this book, I stated that red teamers should use flashlights only when necessary. Light usage must be directed only at the area of concern, should be colored red, and have low lumen output. However, internal lighting is a different animal altogether. An office, for example, is almost always partially illuminated. It is important not to mess with internal lights, but it is critically important to know where these illuminated areas exist.

Avoidance is the best tactic for lit areas. If traversing through it is inevitable, operators must crawl or rush while using surrounding objects for cover. It is strongly advised to refrain from turning off the lights. The sudden change in environment setting could alert someone.

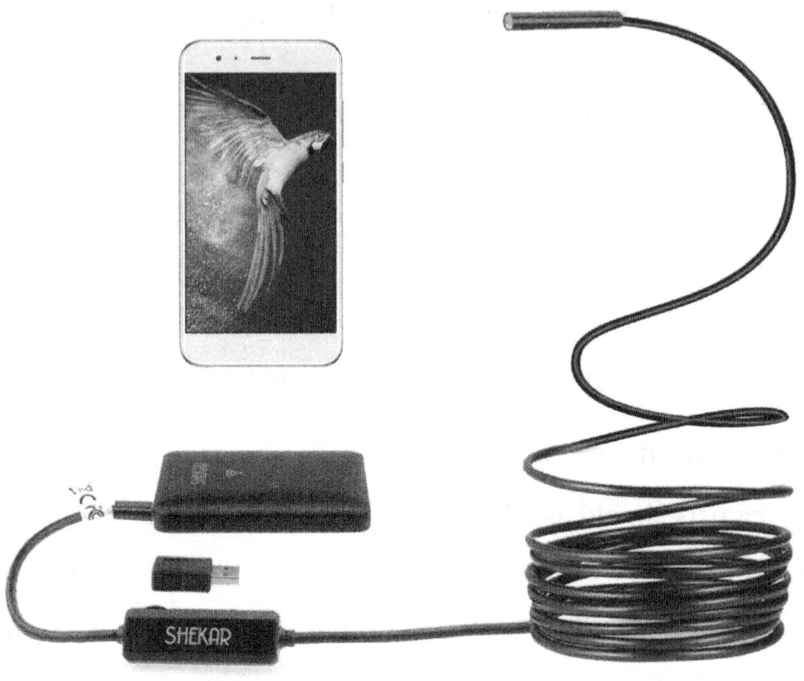

Figure 82. Wi-Fi Borescope

When it comes to doors and corners, a borescope (also called an endoscope) can be used to peer under doors and around corners. In the spirit of light discipline, the model that we use connects to a smartphone via Wi-Fi. This means the light emitting from the phone could compromise our position. But this is one instance where I'm fine with the risk, given the reward. Though I have not used a blue light filter to cover the smartphone screen, I'll bet this, along with turning down the brightness, is enough to mitigate the risk.

It should also be noted that the image a borescope gives is far from high quality and it doesn't do well seeing long distances. However, there's nothing like being able to see inside a room before trying to make entry, even if the image is grainy. A borescope is an ideal tool to check under doors and around corners for security controls (cameras, motion detectors) and avoiding people.

When confronted by an area secured with motion detectors, the first course of action should be avoidance. Find a less secure route. When avoidance is not an option, however, motion detection evasion inside a facility becomes trickier. There is less room to move around and sensors can be, and often are, tuned to levels of higher sensitivity.

Here are three go-to tactics for evading motion detectors:

- Conceal body heat (from PIR)

- Very slow movement

- Angle concealment

I've mentioned how to evade most of today's motion sensors earlier in this book by using a mylar blanket to deflect body heat. The tactic I provided involved fixing the mylar to a wooden frame and building in a handle to prevent hand/finger heat from contaminating the mylar. The idea is no different here, except the wooden frame part.

Evading PIR motion sensors during the Penetrate & Control

phase has to be done with a more limited toolset. Carrying a big mylar blanket and frame isn't going to cut it. Instead, an operator should carry a pair of gloves and fold-up mylar in their tactical bag. The gloves should be used to shield finger heat from contaminating the mylar while the operator holds it in front and away from her body.

Alternatively, a riskier approach is to evade detection simply by moving slowly. Most detectors use heat signatures to baseline a given area's environment. When a sudden change in the heat baseline occurs, the sensor triggers an alarm. Motion detectors are tuned with tolerances for gradual changes that do not abruptly interfere with the heat baseline. Therefore, evasion is possible given the operator moves very, very slowly. Taking 25 minutes to move 15 feet may not jive well with the mission timeline, however.

While this evasive technique is possible, it should probably be used as an option of last resort. I recommend practicing this tactic before considering using it.

A tactic for evading motion detectors and security cameras involves exploiting coverage areas, or lack thereof. Inexperienced security equipment installers mistakenly install security controls too high or create zones of exploitation due to gaps in coverage. As a result, it is possible to evade motion detectors and cameras by slipping through these coverage gaps.

The trick to discovering these coverage gaps is not easy though.

Sensors and cameras installed at sharp angles create vertical zones of exploitation. The same internal sensors and cameras are almost always installed too high far above head height, creating a horizontal zone of exploitation. Operators can successfully exploit these vulnerabilities by crawling, crouch-walking or hugging the wall tightly within the zone of exploitation.

It is difficult to know precisely where exploitation zones exist. For this reason, my team does not leverage this very often. Instead, we would combine this tactic with mylar shielding and slow movement to put better odds in our favor.

Clearing a Room

As the team advances through the facility, they will need to make entry into a room or rooms to reach their mission objective. Sometimes simply making entry into a room, like a server room, satisfies the objective while most of the time they will need to perform a set of tasks like retrieve a piece of equipment, find documents, and so on. Whatever their OPORD might be, the team must do so in an efficient and coordinated fashion.

In physical red teaming, the process of securing the room is called "clearing a room." Unlike law enforcement and the military's use of the phrase, we are not gunning for hostiles. Instead, we are first ensuring the room is suitable for entry. Then we are carrying out our OPORD. As I mentioned earlier, OPORD usually consists of a set of tasks the team needs to perform to successfully complete the mission.

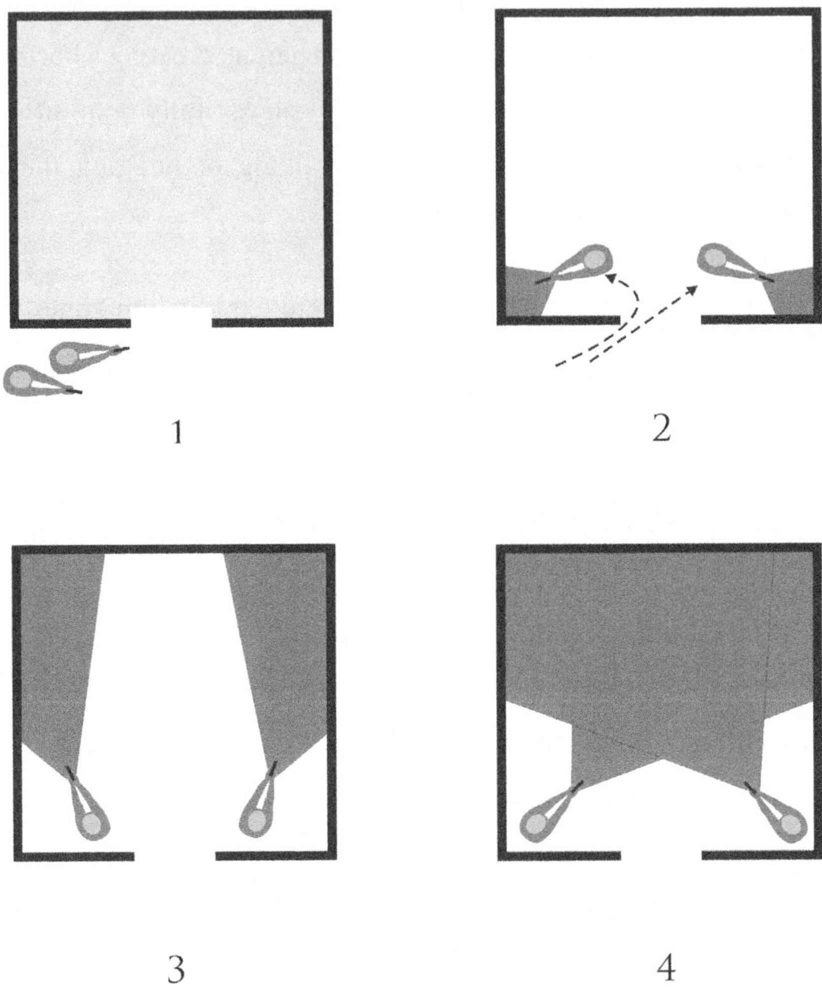

Figure 83. Clearing a Room (guns not necessary!)

A great way to split up clearing a room is depicted in Figure 83. Please excuse the show of guns in the diagram.

Step #1

The team should stack up in a line outside the door. An operator should first ensure entry will not compromise the mission. This usually means checking for security controls and making sure the room isn't occupied. A borescope under the door or other tactic can help with this.

Step #2

An invisible line straight down the middle of the room should act as a dividing line. This is crucial to avoiding doubling up on efforts and wasting time scouring over an area that's already been looked through. Great care should be taken to avoid contaminating the room by pushing papers aside, moving chairs, etc.

Once inside, the team should turn to their immediate corners and begin to clear the room there.

Step #3

As the team continues to clear the room, per the OPORD, it is critical at this point to communicate with each other concerning their findings. It is likely that one (or more) of the operators will have carried out OPORD and this information should be communicated to each other and back to the red team leader.

<u>Step #4</u>

During covert operations, it is vitally important to not leave a trace. Operators must be aware of their body profile at all times to avoid contaminating the room with their presence. So it is important in this step to collect tools and re-situate objects to their original location to reset the environment.

Penetrate & Control sets the pace for movement through a target via controlled entry and progression. Mission objectives are not reachable without considerable protocol, efficiency, and communication at this phase.

[CHAPTER 11]

SECURE OPORD

Several minutes have passed after the crew worker finished his smoke break. The bravo team member who bummed a cigarette from him earlier rapped three times on the external door and the other bravo team member lets him in. The area is partially lit, but free of the cleaning crew. Operational orders in the RoE say they must reach the server room and retrieve an external hard drive left for them by the client stakeholders.

Hugging the darkened wall, the bravo team makes their way toward a pair of French doors they believe leads into the main office corridor. Upon reaching the doors, they rush the hallway stopping every few seconds, still not knowing where they are until one of the team notices a sign saying, IT Department. All enclosed offices and room are centered in the middle of the building. Bravo team rushes each enclosed office/room until one operator notices a room protected by a PIN pad. One operator inserts the borescope under the door. It looks like the right room, but it's dark. The other operator removes the curled up under-the-door tool from his bag and moves it into position. Quietly the operator pops the door latch! They move in quietly, switching to NVGs. It's the server room. Sitting on a server in an open rack is the external hard drive. Kneeling on the raised computer room floor, server fans whirring loudly, cool air circulating and LED lights flashing everywhere, the bravo team radios the red team leader, "Red team leader, this is bravo team. Objective 2 reached!"

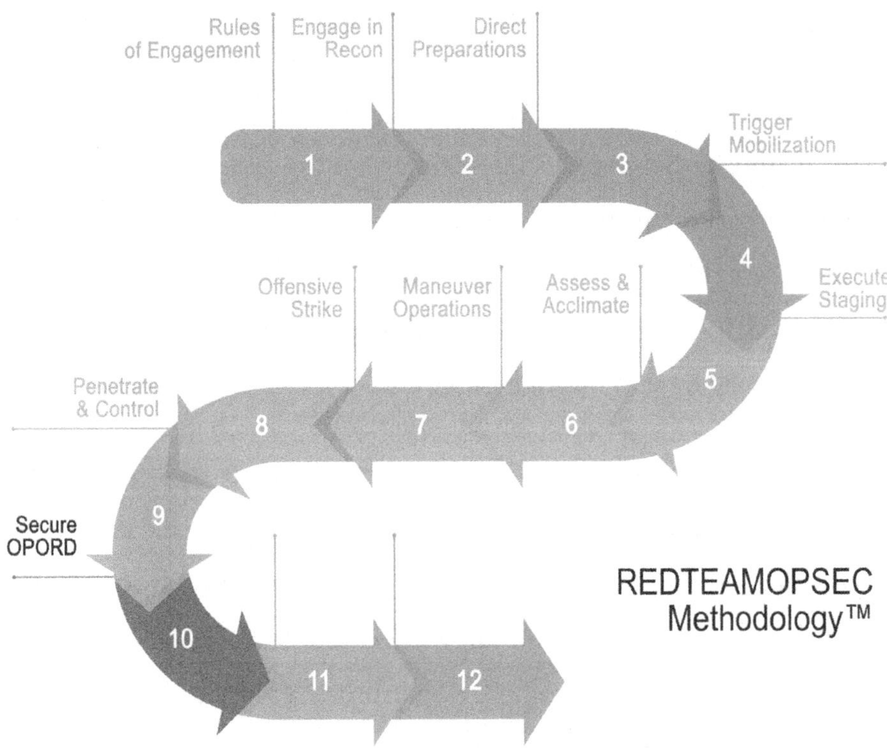

Figure 84. Secure OPORD Phase

Secure OPORD is a significant phase, if not the most significant phase, of the REDTEAMOPSEC methodology, or any given physical red team operation for that matter. It is at this point where the red team carries out the intended actions on objective.

By this time, quite a bit of activity has occurred. Reconnaissance has been performed, the team has staged, moved into offensive position, exploited physical weaknesses, and gained access into the facility. Now comes the time where the all-important objectives are carried out. It's all led up to this.

Execution

Mission Goals							Recon Results		
	Goal	Plan	Mission Success	Est. Vulnerability	Threat	Bad Actor	Observations	Vulnerability	Go-Forward?
GOALS 2	Gain unauthorized access and retrieve a piece of equipment to simulate a physical data breach	Deploy from area #2, approach from east on foot at night. Gain access via unlocked door, change clothes inside for SE. Server room likely at northend of building. Capture video evidence and retrieve external hard drive.	Gain access to facility, find server room, gain access, retrieve an external hard drive and exit unnoticed	Inadequate perimeter security and employee / contractor security awareness	Moderate to Significant Business Impact	National to Regional bad actor. Moderately sophisticated.	Side entrance traffic is used by smokers moderately by day, minimal by night crew. No cameras visible, no motion lights. No RFID badge entry, only door locks that appear to be left unlocked by night crew.	Night cleaning crew leaves door open to take smoke breaks leaving the facility vulnerable to equipment theft, tampering, IP theft, and physical data breaches	Yes

Figure 85. Mission Goal

As I mentioned in an earlier chapter, part of the Execute Staging phase is to reassure every red teamer fully understands what constitutes mission success and how to get there. Figure 85 is taken from planning documents. Inside the Mission Goal is precisely where the OPORD lies and is paramount to the success of the mission.

Let's start on the right of the figure; it shows output from reconnaissance efforts under the heading, Recon Results.

Mission Goals			Recon Results		
		Goal	Observations	Vulnerability	Go-Forward?
GOALS	2	Gain unauthorized access and retrieve a piece of equipment to simulate a physical data breach	Side entrance traffic is used by smokers moderately by day, minimal by night crew. No cameras visible, no motion lights. No RFID badge entry, only door locks that appear to be left unlocked by night crew.	Night cleaning crew leaves door open to take smoke breaks leaving the facility vulnerable to equipment theft, tampering, IP theft, and physical data breaches	Yes

Figure 86. Recon Results (some columns hidden)

Recon Results

This section includes three columns that aim to provide a bit of history from the recon team. This is a helpful reminder to the execution team in the event recon was done by another team. Let's break down this section a little more.

Observations

In this example, the recon team identified a side entrance that is used by smokers, during the day by employees and at night by the cleaning crew. The night cleaning crew had a habit of leaving this door unlocked during their shift. The recon team also stated they did not see any cameras, motion detectors, or RFID readers. We know from the story at the start of a previous chapter that the bravo team found an RFID reader had been installed between phases, throwing a giant wrench into their plan.

Environments are constantly changing. While it may seem to some the client in this case should have told the red team a new security control had been installed, that is almost always not the case. This is partly the reason for the Assess & Acclimate phase, and even then, some things will go unnoticed.

<u>Vulnerability</u>

As a result of the observations the recon team made earlier in the operation, here they indicate what is believed to be a vulnerability. Because the night cleaning crew leaves the door unlocked, it gives rise to the potential for theft of equipment, tampering, intellectual property theft, and overall physical data breaches.

The Observations and Vulnerability columns combined tell a short story about the physical security posture and potential vulnerability identified.

<u>Go-Forward</u>

In the Go-Forward column, the recon team indicates 'Yes' or 'No' on whether this suspected vulnerability is one that deserves to be tested as a mission goal. In our example, the observation here turned into a full-fledged mission goal due to the gravity of the vulnerability.

Mission Goals

Mission Goals		Goal	Plan	Mission Success
GOALS	2	Gain unauthorized access and retrieve a piece of equipment to simulate a physical data breach	Deploy from area #2, approach from east on foot at night. Gain access via unlocked door, change clothes inside for SE. Server room likely at northend of building. Capture video evidence and retrieve external hard drive.	Gain access to facility, find server room, gain access, retrieve an external hard drive and exit unnoticed

Figure 87. Mission Goals (some columns hidden)

Starting from left to right in Figure 87 are the columns most relevant to the execution team. This data is what's most useful during the latter phases of the REDTEAMOPSEC methodology, especially this one.

Goal

"Gain unauthorized access and retrieve a piece of equipment to simulate a physical data breach."

The Goal column provides a brief summary of the objective. I find this data to be most important to client stakeholders. It should be written so that it is easily consumable at a high level. There should be no technical jargon here.

<u>Plan</u>

This column is a summary of the salient action steps the red team will be conducting. It briefly states what, where, and how the team intends to reach the goal as advertised in the Goal column.

From our story about the bravo team, we know there has been some deviation from the plan. We know the RFID reader forced the team to improvise and the plan changed. This is to be expected, and the important thing to keep in mind is that the deviation was not significant enough to be considered out of scope. A character change (clothing) and social engineering were planned, but not in the way the team had originally planned.

<u>Mission Success</u>

The Plan column indicates what, where, and how, but the Mission Success column defines under what circumstances the goal is considered successful. The red team must gain access, retrieve a "flag" (external hard drive), and exit unnoticed. This provides additional parameters on how the goal must be achieved.

"Gain access to facility, find server room, gain access, retrieve an external hard drive, and exit unnoticed."

Plan and Mission Success language should be written with flexibility in it to allow for slight deviations when unexpected issues arise. Even though the bravo team had to improvise on their feet, the spirit of the plan did not change significantly, according to the language. Therefore, mission success for this goal had been reached.

Mission Goals

		Est. Vulnerability	Threat	Bad Actor
GOALS	2	Inadequate perimeter security and employee / contractor security awareness	Moderate to Significant Business Impact	National to Regional bad actor. Moderately sophisticated.

Figure 88. Mission Goals (some columns hidden)

Estimated Vulnerability

Here we more formally and categorically define the vulnerability. As opposed to the Vulnerability column in the Recon Results area, which is more narrative, the language here serves to support what would typically accompany language in a typical security finding.

Categories like this are used to better organize findings that are subject to remediation efforts and also to help clients understand which categories need the most improvement.

Threat

The Threat column is relatively straightforward and brief. This communicates to clients the perceived threat to the organization as a whole.

Bad Actor

This column indicates how sophisticated a would-be attacker would need to be in executing this goal successfully. It also indicates if the bad actor is more likely to be local, regional, national, or international.

Again, the OPORD provided in the Mission Goal is at the heart of the operation and must be fully understood in order to effectively reach mission success.

SITREP

We know from earlier in this book that a SITREP is a situational report used to notify senior-level leaders of a tactical situation and status usually occurring after a significant event. It is a short and concise statement and, in a physical red team operation, is usually broadcast over a two-way radio to the red team leader. The red team leader may request a SITREP on an ad-hoc basis as well.

Figure 89. Me receiving a SITREP (Credit: Paul Szoldra/Tech Insider)

The red team leader uses the information relayed in a SITREP throughout an engagement to make critical decisions regarding the operation as a whole. However, the SITREP following an operator having reached a mission goal during this phase, as in the story at the start of this chapter, is the pinnacle of an operation.

It is at this point, and this point only, after having received a SITREP that the red team leader, in her position of operational power, declares mission success, mission failure, or otherwise. But before we get into the specifics of other outcomes, allow me to provide a simple SITREP template for operators to use.

There is no requirement to adopt military radio etiquette. The SITREP template here is merely an example a team can use to efficiently communicate. At such a pivotal point in the operation, this is one area where your team may want to utilize military radio etiquette if you have not already done so.

Mission Standing

When things go right, the red team leader will announce mission success, in response to the SITREP, and order the team to exit the target. It is usually evident when a mission has successfully reached its objectives. The red teamers know it, and the red team leader gives the order to exit. But things don't always go as planned. Sometimes mission objectives aren't reached or sometimes they take longer than expected. Sometimes operators are compromised, and mission success turns into mission failure.

Here are a few red team leader mission orders:

- **Resume OPORD.** The red team leader grants additional time for the team to attempt to secure OPORD.

- **Concede.** For example, if an operator is compromised by an employee, the red team leader will give the order to show authorization letters, ID, and forfeit the mission.

- **Abort.** The red team leader believes the objective can't be reached and orders the team to move on or exit the target entirely.

It is important for the team to know the meaning of these orders and what to do. The most common of these for my team is resuming OPORD. Chance, randomness, and surprises tend to pop up more than we plan for, and it seems mission goals take longer than we anticipate. Operators should provide SITREPs to the red team leader when encountered with unplanned issues that delay the OPORD process.

Conceding often happens when operators are spotted by employees, guards, or law enforcement.

In the case of employees, operators should always attempt to social engineer their way out of a sticky situation. A pretext and character change, as I mentioned in Chapter 11, should be used to further those efforts. Conceding to employees should only be done when it appears that no way out is possible.

When confronted with law enforcement, on the other hand, the process should be handled much differently. Any and all operators must immediately concede in the event they are compromised by an officer. This means providing authorization letters, government-issued ID, and complying with the officer's commands. Do not lie, run, or hide from law enforcement. I repeat, do not lie, run, or hide from law enforcement. You could be putting yourself into grave danger, let alone serious legal ramifications.

Aborting a mission goal is prone to happen and it shouldn't be frowned upon. It's probably a sign that the client is doing things right. In a physical red team operation, there are almost always a multitude of mission goals. Not every goal can be or is expected to be reached. Aborting all mission goals, on the other hand, is a different story altogether.

I imagine there are a few scenarios where aborting the entire operation is necessary, though I've never been a part of one that has. However, physical injury and law enforcement intervention fall well into that category

276

[CHAPTER 12]

EVACUATE, EVADE, & COVER

A member of the alpha team radios, "Objective #1 reached" into the handset. Now that both teams have communicated that their objectives have been reached, the red team leader gives the order to exit the target. The alpha team, having completed their goal of making entry into an executive's office, quickly blazes a trail back to the warehouse and through the maze of pallets. Just as they exit the loading dock door, a pair of headlights briefly wash over the entrance, and the team runs behind a dumpster. With headlights now squarely positioned on the dumpster, the vehicle isn't moving. They wonder if they've been spotted.

With hard drive in tow, the bravo team resets the server room to its original state and heads back to the employee entrance. Thankfully, the hallways are still dimly lit, and they can see the lights in the same room they entered through. Rushing down the hallway, almost to the room, they hear voices. One bravo team member runs inside the room, meanwhile the street-clothed member circles back to hide in the empty break room. The voices are getting louder near the break room. Thinking quickly, the street-clothed member turns on the break room light. Voices getting even louder, he peeks around the corner to face two night crew cleaners. "Hey, do you have change for a five? This vending machine only takes \$1 bills." The workers are startled at first, but believing he is an employee working late, they apologize for not having change and continue down the hall. Waiting a moment, the last member of the bravo team makes his way back down the dimly light hallway, catches up with the other team member, and they exit through the employee entrance.

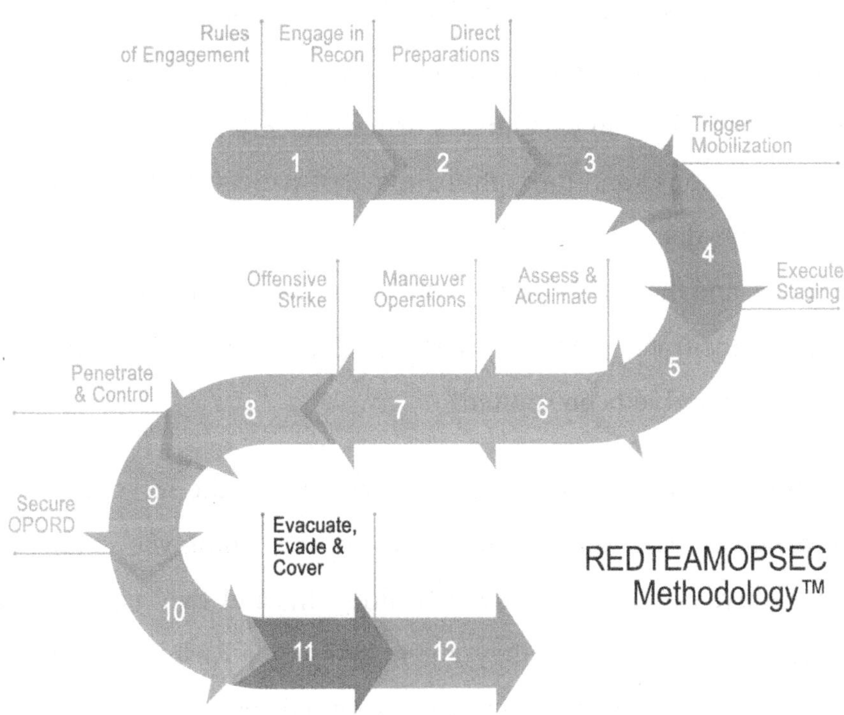

Figure 90. Evacuate, Evade, & Cover Phase

Evacuate, Evade, & Cover is a small and often overlooked stage. Once a mission objective has been reached, one would think it would be time to just make a run for it. On the contrary, there are many things to take into consideration. Where is the team going to exit? Where are they going to rally at? What or who should they avoid on the way out? How should they cover their tracks so no one suspects anything after they leave?

Consider that most physical red team engagements expect operators to exit the facility in order to consider the operation fully executed. Sometimes this means a full covert operation by which none of the operators are to be discovered. Other times, it may call for social engineering tactics, similar to what the bravo team used in our story. Either way, a full and clean exit is usually compulsory. This chapter aims to help explain that better.

Evacuate

First off, let me explain why the term evacuate was used to describe this step when it could have simply been titled exit. Think of an evacuation as more of an orderly exit from a facility. I want to underscore the significance of an orderly exit.

Once the bravo team nabbed the external hard drive and reset the server room, they made like a wild banshee and rushed for the door. It's natural to want to flee precariously after having done something "bad." However, just as carefully as you infiltrate a facility, you must evacuate it just as carefully.

It's easy to expect hallways and rooms to remain unchanged, even if you just passed them seconds ago. Expect the environment to change. Constantly. Otherwise, the lapse in situational awareness will surely lead to grave mistakes.

Building Layout

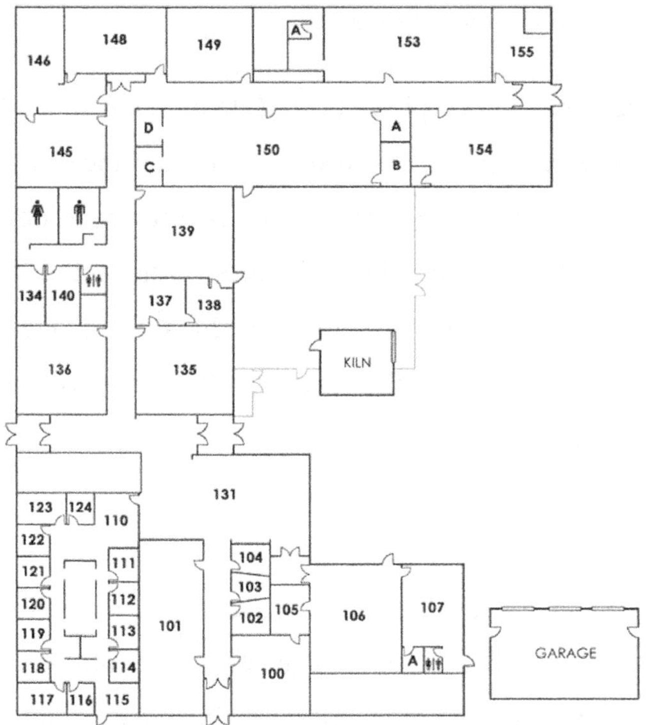

Figure 91. Building Map

Identifying building maps to help navigate in and around a facility are often discovered during the Penetrate & Control stage. Then, the goal was to find certain rooms to gain access into. At this stage, operators should use information in building maps with the intent of looking for alternate exits and additional areas to avoid. There is no hard requirement that a team has to exit the same way they entered. Though most operations involve exiting through the same entrance, finding an alternate exit closer to the rally point may be ideal. In any situation, building maps should be sought out at this stage in order to enable an orderly exit from the facility.

Rally Point

Figure 92. Rally Point

Critical to the next step of the evacuation stage is the rally point. The rally point is the designated location outside the facility where red teamers will meet once ordered by the red team leader. Recall from our earlier chapters that the red team leader gives the order to rally once all mission goals are met. The rally point is where the red team leader waits at a nearby location, usually in a vehicle, during the execution process. All red teamers must know exactly where to find the rally point. As stated earlier, it is their sole destination once their mission goals have been met.

Figure 93. Staging, Deployment, and Rally Points

A thorough study of the operation's staging, deployment, and rally points is necessary during the operation. Optionally, an additional rally point called an emergency rally point could be created. If the facility is hot with lots of personnel onsite or the operation involves several mission goals, an emergency rally point could provide a temporary safe haven during the engagement.

There are unfortunate situations where the red team leader orders the team to the rally point when things go sour. The reasons are varied, but usually, because an operator gave it their best, but for some reason couldn't complete their goal successfully. Difficulty exploiting vulnerabilities are the usual suspect. This does happen and should not be frowned upon. Unless there is language in the SOW that says otherwise, a team could make another attempt where it seems reasonable and realistic.

Evade

Evasion is defined as an act of escaping, avoiding, or a trick to get around something. We accomplished this during infiltration, but we were equally focused on trying to find unknown security controls. During evacuation, we are keenly focused on evading those security controls and people. To our benefit, we now have a better sense of the in-place security controls and how to avoid them and the movement of people, if any. As a result, this usually enables us to move quickly during the evacuation.

Movement during the evacuation involves at least two positions. The first is a crouched walking position enabling moderate movement inside a target. The second is the dash. Dashing is utilized once the red team is outside the facility and en route to the rally point.

Dash Movement

Figure 94. Example of the dash movement

Similar to the rush movement in the Maneuver Operations chapter, the dash is the fastest method to get from one point to another. Similar to rushing, dashing starts from a dropped position and moves to a crouched sprint. Sprinting is what makes this movement different from rushing. Dashing about every five seconds is ideal since it makes it difficult for potential bystanders to fully track movement.

Just as with rushing, each dash must have a designated drop destination. In other words, an operator should not drop to the ground simply because five seconds have passed. A drop destination providing some level of cover or concealment should be sought out in conjunction. Optimally, the drop should include the time to cycle through an OODA loop.

OODA

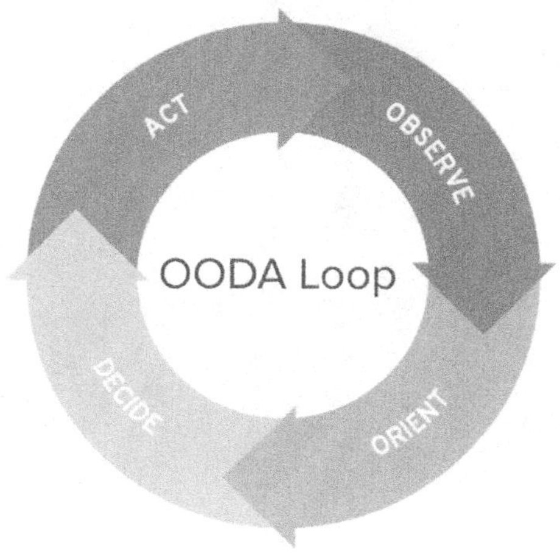

Figure 95. OODA Loop

Recall from our earlier chapter, "Maneuver Operations," a description of an OODA loop. It stands for Observe–Orient–Decide–Act and is a decision-making framework used to train soldiers to make decisions when there's little to no time to assemble all the data. The OODA loop framework enables red teamers to filter available information, establish context, and quickly make the best decision using the information available at the time.

Red teamers must continually observe, orient, decide, and act as they encounter and evade new obstacles during the evac process. A facility's environment is always changing, and operators must know how to conduct OODA loops to support their operation.

Figure 96. Evasive maneuvers

There are all sorts of obstacles that could potentially give away the position of a red teamer and make it difficult to reach the rally point smoothly. This stage is all about evasion, and evasion is all about being covert. That means going unseen and unheard wherever you are. However, in the thick of an operation, it is difficult to be situationally aware within 360° of your body, and it's easy to walk into a hairy situation.

When evacuating and evading, here are a few key strategies to use:

- Chart an exit path to the rally point with the most shadows

- Be mindful how your profile casts shadows dynamically as you pass different perspectives of light

- Become situationally aware of the sound of your movement (i.e., footsteps, radio chatter, tools)

- Make heavy use of cover (i.e., hug shadowed walls and avoid lit areas)

- Tune in to the sounds of the environment (i.e., traffic, voices)

- Peek before moving through doors, around corners, and using stairs

- Radio your teammates and warn them of any obstacles

Leveraging these key evasion strategies will help enable a smooth evacuation. Remember, evasion is all about quick dash movements, OODA loops, and becoming hyper aware of your surroundings and your body profile.

Cover

Figure 97. Covering physical evidence

Covering tracks is an effort to hide or destroy evidence of the physical red team's actions and presence. There are usually many physical targets to a single operation and failing to adequately cover physical evidence may put the target on high alert. You wind up contaminating the test environment since employees react much differently if they become highly suspicious. Unless this is the intent, red teamers must carefully cover their tracks.

Let's start by further defining what it means to cover your tracks. Yes, to a certain degree it does involve actually obscuring your footprints in applicable environments in dirty and rugged terrain.

However, it's meaning is much more than that.

Office

Figure 98. Me scouring through an office in an actual red team operation

Red team operations that require operators to scrub offices and cubicles for sensitive documents are fairly common. These areas must be treated with the utmost care so as not to leave a trace. I'm always careful when I need to move keyboards, chairs, mice, phones, coffee mugs, monitors, and laptops. Anything the user touches throughout their daily routine that is even slightly out of place will certainly raise an alarm. These personal workspaces and objects are sacred ground to many office workers, and they can tell when something is wrong.

I always look under keyboards for passwords and other sensitive information. More often than not, I am rewarded with something useful. When peering under keyboards, I use two hands to lift up the side that is closest to me giving just barely enough room to look under. It's as if the keyboard is attached to the desk with a hinge on one side. You should never lift a keyboard or any object completely off the desk, unless absolutely necessary. Peer under it instead. This one tactic alone will go a long way in minimizing your physical presence.

If keyboards are the most common object I handle in an office, the second object would be the office chair. When office workers leave for the day, they usually swivel around and get up from their chair to exit. This position places the seat of the chair outward, ready for the worker to sit down. An office worker's chair that is out of place is one of the first things they notice. They will know someone has been there. When re-positioning a chair, take a mental picture to be certain it is put back in its original place.

During some of my previous covert military training, the instructor recommended taking a before and after photo as an insurance policy. There are smartphone apps on the market that make before and after comparisons even easier. Offices are chock full of disasters waiting to happen, such as accidently knocking over a pile of papers, etc. So there is reasonable cause to resort to taking before and after photos. Although I have never done this in the field, I do see the benefits. However, I cannot caution those considering this tactic enough to turn off the camera's flash and volume.

Lights

Figure 99. A well-lit office building

There is a general rule about lights in the world of physical red teaming. If the lights are on, leave them on. If the lights are off, leave them off. That said, my team has violated this rule in the field under very specific circumstances.

However, let it be known that altering the environment in such a drastic and visible way, such as turning on the lights, is a very risky move; a decision to be made after carefully weighing the circumstances. But because so many of the red teams I train fail in this department, I felt it should be covered here. Stick to the general rule and leave the lights where they are.

Locks

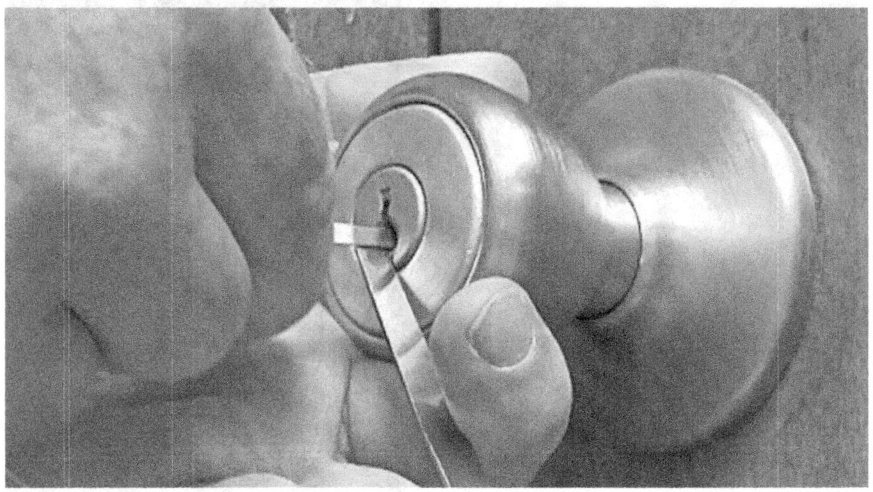

Figure 100. Picking a lock

If your physical red team operation involves lock picking, you could be leaving behind one very significant clue. So you picked a lock open to the networking closet, achieved your mission goal, shut the door behind you and evacuated. If you didn't pick the lock closed, you could be in big trouble.

By exploiting the manufacturing defects in the lock itself, you've managed to pick it open with your pick set, not a key. The lock will remain in its open state until you pick it closed, aka locked. Remember, you don't have the key, so you'll need to use your pick set to relock it. If the lock is protecting something critical, like access to a restricted area, this may result in the filing of a formal incident report and could put the target on high alert.

Terrain

Figure 101. Overshoes made of carpet to hide tracks

During red team operations, our goal is to become covert, and hiding footprints in various forms of terrain and weather is very challenging. One of the best ways to cover physical tracks, like shoeprints, is to use overshoes. In fact, one of the most effective ways to obscure boot prints is to use overshoes made of carpet. The thick fabric obfuscates the tread and, in some cases, hides the foot impression altogether. You won't find these at the store, you'll have to fashion these by hand from carpet remnants. I highly recommend these for use in dry to moderately wet ground.

A different set of overshoes is needed to cover up prints inside a facility.

Figure 102. Overshoes big enough to fit over boots

High top overshoes, like those pictured here, are ideal to avoid leaving a trail of moisture or mud from the rough terrain to sidewalks, entrances, and hallways of a facility. Some facilities have surrounding terrain that would make a very noticeable mess, not to mention leaving a telling trail of evidence. Rubber overshoes work the best in keeping what's underneath dry.

A pair that can be swapped out quickly and stored inside a tactical bag is even better.

Moist and muddy ground is the antithesis of what you should be traversing. Avoidance is the best tactic, but this is not always possible. You want to seek out terrain of hard-packed soil or rocks and pebbles. This terrain is the best to use so as not to leave any visible tracks.

Snowy ground is another animal altogether. I find this the most difficult environment to traverse covertly. I suppose operators could use snowshoes for more sophisticated operations, but that isn't realistic for most operations. Covering the tracks for a team of operators is nearly impossible. The best tactic in this situation is to obscure the size of the team to look like one individual, instead of multiple. This is done by having the entire team walk in single file while stepping into each other's snow prints. The idea is that anyone noticing the prints won't be as concerned since it merely looks like one person was walking about instead of a multitude of people.

For hands-on Physical Red Team Training, please visit:

https://www.redteamsecuritytraining.com/physical-red-teaming-book

[CHAPTER 13]

COLLECT & EXFILTRATE

The alpha team radios the red team leader, "I think we've been made. We're behind a dumpster outside the entrance, but a car has its lights on us, and we can't move." Then, the engine and headlights stop. Two men exit the car and enter through the loading dock door. The alpha team is relieved they weren't spotted. The men were dressed as the cleaning crew. They must be back from a late-night break. The alpha team radios a SITREP to the red team leader and she orders them to proceed to the rally point. As they move out, one operator remarks quietly to the other about how noticeably active the location is compared to the recon phase.

The bravo team, meanwhile, had nearly reached the waiting van at the rally point until the red team leader radioed for all teams to hunker down. Just in front of the van, a few people walking down the sidewalk stopped to talk and were soon joined by a car who pulled up beside them. The bravo team drops to the ground a mere thirty feet away. The alpha team catches up with bravo team and drops down next to them. Both alpha team and bravo team have collected at the rally point but are unable to enter the van without being seen.

Realizing the bystanders wouldn't be going anywhere soon, the red team leader calls for the operators to low crawl to the rear of the van facing away from the small crowd. One by one they quietly crawl to the van and make entry through an open door to the back. Once the team is collected, they perform a quick equipment check to make sure nothing has been left behind. All operators crouch down out of sight of the van's windows as the red team leader fires up the van and they calmly pass the crowd and exfiltrate the target.

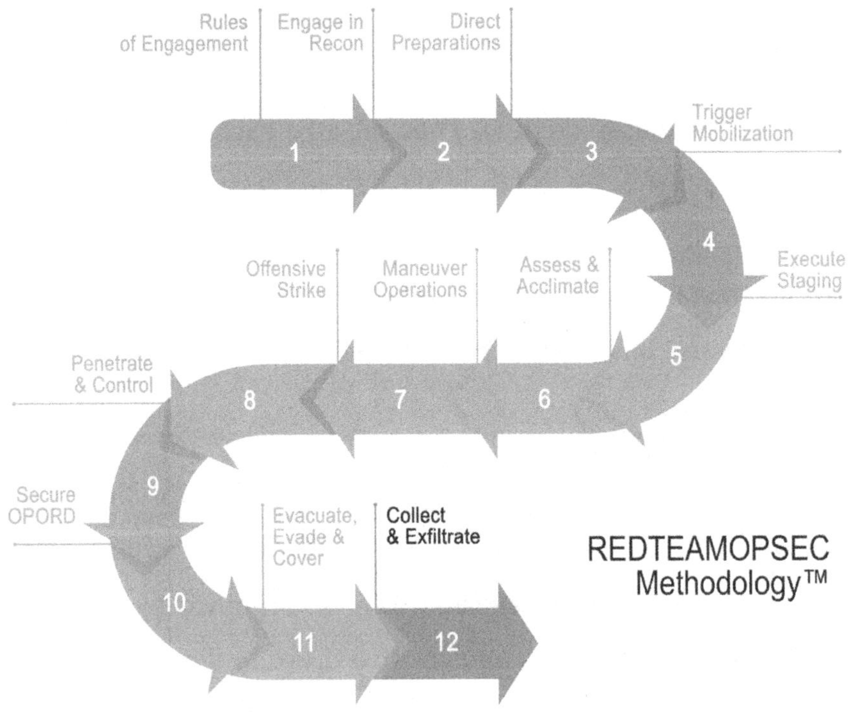

Figure 103. Collect & Exfiltrate Phase

The story at the beginning of this chapter is not all that far off from what happens during actual operations. What is important to take into account is that exiting a target is more than just walking out. Exfil takes planning, and it takes skill when things go sideways. Remember that most physical red team operations are not considered fully executed until the entire team exits the target cleanly. That's when the red team leader hits the gas in the getaway van with the entire team in tow.

A proper exit that supports a fully executed red team operation is one that happens in two generic stages, collect and exfiltrate. Let's take a closer look.

Collect

There are so many opportunities to accidentally leave behind equipment at the target and misplace or forget evidence altogether during an operation. I should know. I've made these mistakes myself, and it nearly cost me the operation. I was gloating a little back at the hotel after completing what I thought was a successful mission inside a secure substation yard. Then a shockwave hit me when I realized I didn't have my Shove-it tool. I looked everywhere for it. After some thought, I was pretty certain I dropped it after climbing over one of four barbed-wire fences. In fact, I was even more certain it was probably inside the inner-most fence closest to the building we broke into, along a path heavily used by its employees. It was not a proud moment.

In this stage, I hope to provide guidance on how to properly collect evidence, equipment, and rally with fellow operators for a successful mission.

Evidence

Figure 104. An example of a bag to collect evidence

A fair percentage of physical red team operations I have been privy to have included the capture of physical evidence. This is unlike a flag, which is a designated object an operator sets out to acquire.

Figure 105. Red teamer capturing evidence during an infiltration

Evidence might be a sticky note with root credentials, a document with confidential information, or sometimes an untethered laptop. Evidence is a security risk an operator happens upon by chance that she finds during execution and is found to be relevant to the nature of the mission objective. For example, an untethered laptop would be taken as evidence if the company's concern is theft. Confidential documents would be taken if the company is concerned with unauthorized data disclosure. Of course, how this evidence is captured is entirely dependent upon the RoE. But it is fair to say that most often the evidence will be physically taken with the operator as opposed to only being photographed.

In short, red teamers in an operation that allows for the physical acquisition of evidence must plan for it by having adequate storage on their person to accommodate the evidence they find.

Equipment

My story about losing a piece of equipment is a real threat to red teamers, and it has dire consequences. Recall from the "Execute Staging" chapter where I advised red teamers to securely pack their gear to prevent it from falling out. This advice applies to every time an operator pulls a tool from their pack and replaces it.

An equipment check should be conducted at this point to prevent leaving something behind and raising alarms by anyone who sees it. In my previous example, I left behind a Shove-it tool. This piece of equipment looks a lot like a Slim Jim used in years past to break into cars. A giant wrench would have been thrown into our entire operation had anyone noticed it laying around near the entrance.

To reduce the chances of accidentally leaving equipment behind, I recommend making a cheat sheet of packed gear and where each piece of equipment is held.

PACKED GEAR						
		Large Comparment	Front Pouch	Left Pouch	Right Pouch	Bottom	Top Zipper
BAG	Tactical Backpack	Under-the-door-tool	Laptop	LED headlamp	PlugBot, Shove-it, pick set	Small torch,	USB drive

Figure 106. Packed gear cheat sheet

I almost always use the same tactical backpack and almost always put the same pieces of gear into the same compartments. But even as a seasoned red teamer, I know the anxiety that comes about in the midst of an operation, and that will cause mistakes. A small printed copy of the cheat sheet is useful in ensuring pieces of equipment are not accidentally left behind.

Operators

This brief but necessary step is more about communication than anything. It's imperative all operators know it is time to make for the rally point, to what location if it has changed, and if there are any hazards to consider. From the story earlier in this chapter, the red team leader gave the authorization for the alpha team to head to the rally point. Later in the story, the red team leader communicated concerns about a crowd gathering near the rally point. As a result, a slight deviation from the original plan was necessary in order to make a clean exit.

Essentially, this step is here to ensure the operators are able to collect at the rally point safely and make a clean exit. The onus for this step falls mostly on the red team leader. But it is important for the red teamers to communicate and coordinate similarly to support a fully executed operation.

Exfiltrate

The term exfiltrate is defined as the process of withdrawing from a place or stealing sensitive information from a computer. As you might've already guessed, the REDTEAMOPSEC methodology applies directly to the physical withdraw from a place. But there is a fuzzy line between physical red teaming and the exfiltration of electronic data.

Unfortunately, many physical red team operations today do not cross into the cyber realm to also include ethical hacking tactics. Testing is often compartmentalized to physical security without regard for how physical security vulnerabilities also impact cyber vulnerabilities and personnel vulnerabilities. To combat this, my company RedTeam Security created an approach called Full-Force Red Teaming. This is a more complex issue to be covered in this chapter. For the sake of brevity, please refer to the final section of this book titled, "Full-Force Red Teaming."

Flags

I've used the term flag throughout this book. Let me take a moment to expand on it some. The term flag is derived from a game called Capture the Flag (CTF). It is a traditional outdoor game where two teams each have a flag (or another marker), and the objective is to capture the other team's flag, located at the team's "base," and bring it safely back to their own base.

As you can see, our use of the term flag is loosely based upon capturing an object from our client and the similarities tend to end there. A flag can be anything and usually ranges from a piece of old equipment to a physical document and everything in between.

Figure 107. Capturing a flag in a data center

Operators will certainly know which flags need to be captured and must plan for it accordingly. Keeping track of them during an operation is usually not difficult since there are generally only one to three flags per target. Operators can create a cheat sheet, similar to one used for packed equipment, to better manage flags captured if necessary.

Rally Point

Figure 108. Example of a rally point

The rally point is the final destination for red teamers once their mission objectives have been met or for other reasons as deemed necessary by the red team leader. For most teams, the rally point is a location near the target but far enough away to go unnoticed from employees and most casual passersby.

This is the last leg of the operation. The red team leader coordinates the effective and efficient exfiltration of the team, ideally leaving nothing and no one behind.

Figure 109. Red team leader at the rally point

Once all red teamers are at the rally point and presumably inside the vehicle, the red team leader must take time to conduct the following before declaring mission completion. Please have a look at the following considerations.

Considerations before declaring mission success:

- Roll call. Are all red teamers present?

- Do red teamers have all of their equipment? Was anything left behind?

- Can red teamers re-attest they've successfully accomplished their mission objectives?

- Are flags and evidence captured present?

If these criteria are met to the red team leader's satisfaction, the mission can be considered complete. At this point in time, a proper team exfil is in order and the team can now physically leave the premises.

The red team leader must immediately communicate the status of the operation to the designated stakeholders by phone, email, or face-to-face, whichever communication medium was agreed upon.

Figure 110. Mission complete

With the whole team in the vehicle, it's usually right about now when the air becomes filled with the sound of enthusiastic conversations and stories of close calls. Before too much time passes, I highly recommend all operators scribble notes about the operation, noting important times and events. This should happen immediately, even while still in the vehicle. As the minutes and hours wear on, details start to become fuzzy. So it's important to capture these tidbits before they're gone from memory.

Team Debrief

At the earliest convenience, the team should meet to momentarily debrief and document a rough timeline of events. I suggest constructing the timeline on the same night or day. I'll admit this is one of my least favorite things to do in the middle of the night after a long day. However, specific details concerning times and events are usually the first to escape from everyone's memory, so it's best to do it as soon as possible.

When the team is rested and more productive, a proper debrief should be conducted and the timeline should be formalized. I suggest the red team leader should interview each team member to support the development of the timeline. During this process, all operators should turn in their videos/photos to the red team leader along with any flags or evidence captured.

The debrief efforts will ultimately support the development of an initial draft of the physical red team operation report. My hope is the

REDTEAMOPSEC methodology is one that will prove to be useful to you and your team in all aspects of physical red team operations.

312

For hands-on Physical Red Team Training, please visit:

https://www.redteamsecuritytraining.com/physical-red-teaming-book

[CHAPTER 14]

FULL-FORCE RED TEAMING

Offensive security testing has developed and grown over the past few decades, and we have advances in our nation's military to thank for it. In the private sector, systems penetration testing arrived mid-century. Social engineering testing arrived not long after. And for many, physical red teaming is the new kid on the block. Certainly no one can be blamed for the staggered evolution of offensive security testing. However, it has affected how the three testing strategies, Physical Security Testing, Social Engineering, and Technical Penetration Testing, are used in the private sector.

Throughout this chapter, I will refer to the three testing strategies, physical security testing, (technical) penetration testing, and social engineering as the **Security Testing Triad**.

Please see Figure 111, the Security Testing Triad.

Security Testing Triad

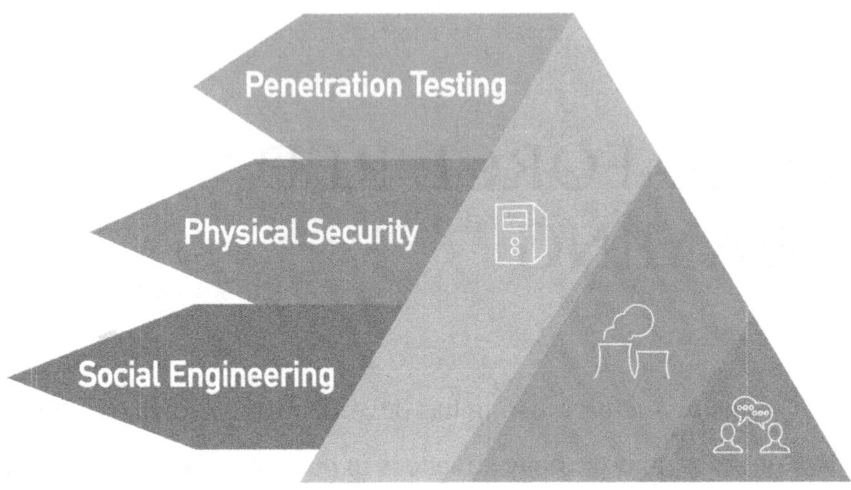

Figure 111. The Security Testing Triad

Synopsis

Technical penetration testing, social engineering, and physical security testing make up the three domains of the gold standard for the offensive security testing triad. I can say with a great deal of confidence that in the real world they are rarely ever integrated together during an operation. What's more, the vulnerabilities discovered from one domain are never carried over to other domains and analyzed further in an effort to determine their collective impact and likelihood for more serious concerns. This is the textbook definition of **siloed** testing.

Siloed testing limits the scope of testing and it has a detrimental impact on the state of information security within a given organization.

Full-Force Red Teaming® is the direct result of a patent-pending offensive security testing methodology I developed with the intent to address these grave issues.

To explain in greater detail, consider the following sections.

Problem

When offensive security testing is performed, usually technical penetration testing, it is often conducted without regard for how the vulnerabilities discovered during the engagement could impact the other domains, physical and social collectively. Failure to test all three domains comprehensively and mutually with each other's results is an overall failure to measure *actual* security risk.

Let's take a moment to spell out the primary problems:

1. The shared mindset of many red teamers, while adversarial in nature, concentrates primarily on applications, networks, and systems only.

2. Social and physical risks are not discovered, analyzed, or addressed. Offensive security testing is not performed comprehensively to include technical penetration testing, social engineering, and physical penetration testing.

3. Findings within domains are calculated irrelevant to one another, as if they do not impact each other.

As stated earlier, many red teamers focus entirely on testing applications, networks, systems, etc. Now, they are approaching this in an adversarial style, as red teaming strategy infers, but they are only addressing the technical bits. In other words, physical and social weaknesses are not discovered, evaluated, correlated, and dealt with during these red team operations. Unfortunately, the measured outcome of such uncomprehensive testing is mistakenly adopted as an accurate metric for the overall security posture of the organization.

As you can imagine, incomplete testing leads to all sorts of untruths that snowball into misinformation. What's more, the findings that are discovered during testing, usually technical penetration testing, are not correlated or linked to their impact on physical or social risks.

To further explain, the diagram shown in Figure 112 illustrates some of the major problems with the current state of the union concerning red teaming.

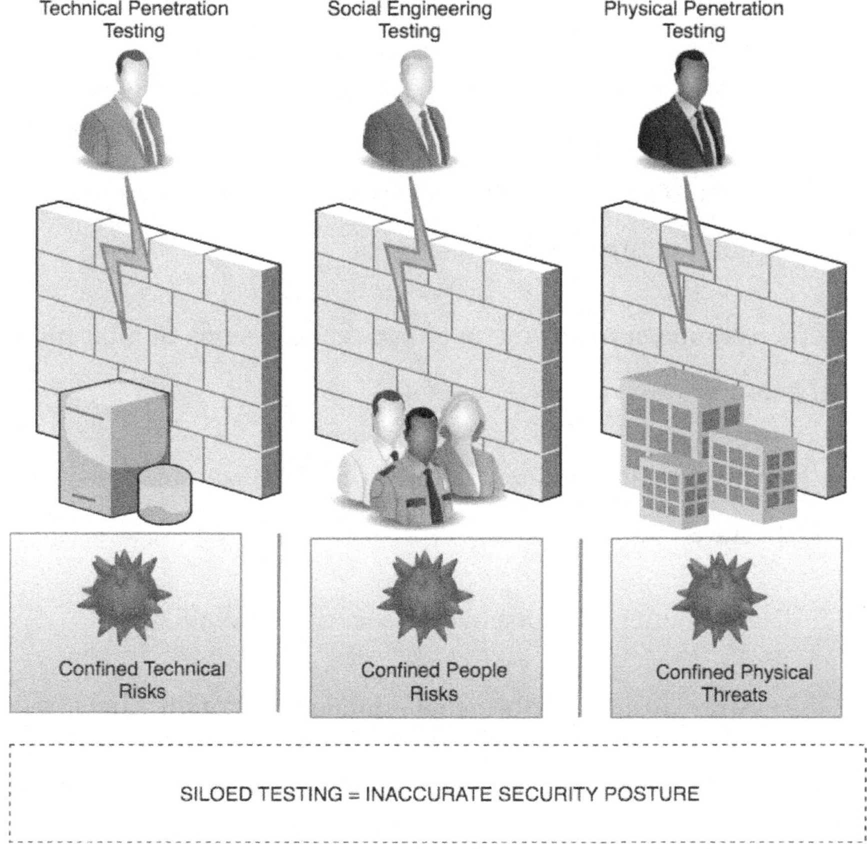

Figure 112. Compartmentalized, or siloed, security testing

Siloed testing, or security testing performed in a vacuum, as shown in the illustration, leads to a number of issues. The following section points out some of the primary problems that siloed security testing can have on a given organization.

Result

Taking into consideration the aforementioned problems with today's implementation of red team testing, we know that this approach fails to measure *actual* security risk. The adverse effects, however, don't entirely end there.

Let's examine some of the negative repercussions in a bit more detail below:

1. Critically inaccurate picture of organizational security posture

2. New frontier of undiscovered security weaknesses

3. Testing is not performed commensurate with client risk profile

4. Misappropriation of costly budget dollars

5. Greater risk for data breaches

6. False sense of security

The absence of comprehensive testing to address the three domains of offensive security paints a critically inaccurate picture of an organization's security posture. What an organization believes to be secure in one domain does not relate to how other risks in other domains have an impact. This is true even if an organization tests all domains, as they almost certainly test them in a vacuum independent of one another, aka siloed testing.

Siloed testing does not account for interrelated risks across domains and can result in a brand-new frontier of undiscovered and unchecked security vulnerabilities. Low or insignificant risks in one domain may bring about grave risk in a related domain or two. It's extremely common to see blaring physical and personnel security weaknesses provide an easy compromise around hardened network perimeters simply because testing and evaluation are not performed inter-relationally.

Network perimeters are often the most hardened. They have been given far more attention over the years and have received far more security budget dollars. Why? It has been the belief for many years that a security breach will surely originate from the network perimeter. This may have been true in the past, but it is no longer the case, and this is precisely where many organizations get stuck. Precious security budget dollars end up reinforcing only the network perimeter, over and over again, while turning a blind eye to physical and social risks.

For the few that actually carry out physical security testing, TTPs and complexity of the planned testing is often lopsided with the real-world threats the target or client is likely to face. Most physical ops teams do not accurately model risks and tailor their tactics to be commensurate with the complexity and type of probable attack vectors. Performing a sound physical security op testing the "right" things in a commensurate fashion is often lost to a lack of risk modeling, or it is the desire to feel like James Bond or Ethan Hunt? I call this the Mission Impossible Effect. For example, a team of eager red teamers using $100,000 in gear while utilizing Mission Impossible style tactics on a mom and pop pizza shop whose assets total only $10,000; there is nothing learned or gained, on either side, when an operation is not commensurate or realistic.

It is clear that an incomplete view of the security threat landscape can lead to gross misrepresentation and misunderstanding of the security posture of an organization. A heavily fortified network perimeter will do little, if anything, to stop physical or social engineering attacks. It seems obvious when you think about it. But all too often, I have witnessed many clients realize this cold and bitter fact in real-time. Ripping off the warm cozy blanket of false-sense-of-security is not something I enjoy doing.

The following section addresses most of these concerns through Full-Force Red Teaming. Take a moment to consider the illustration in Figure 113.

Solution

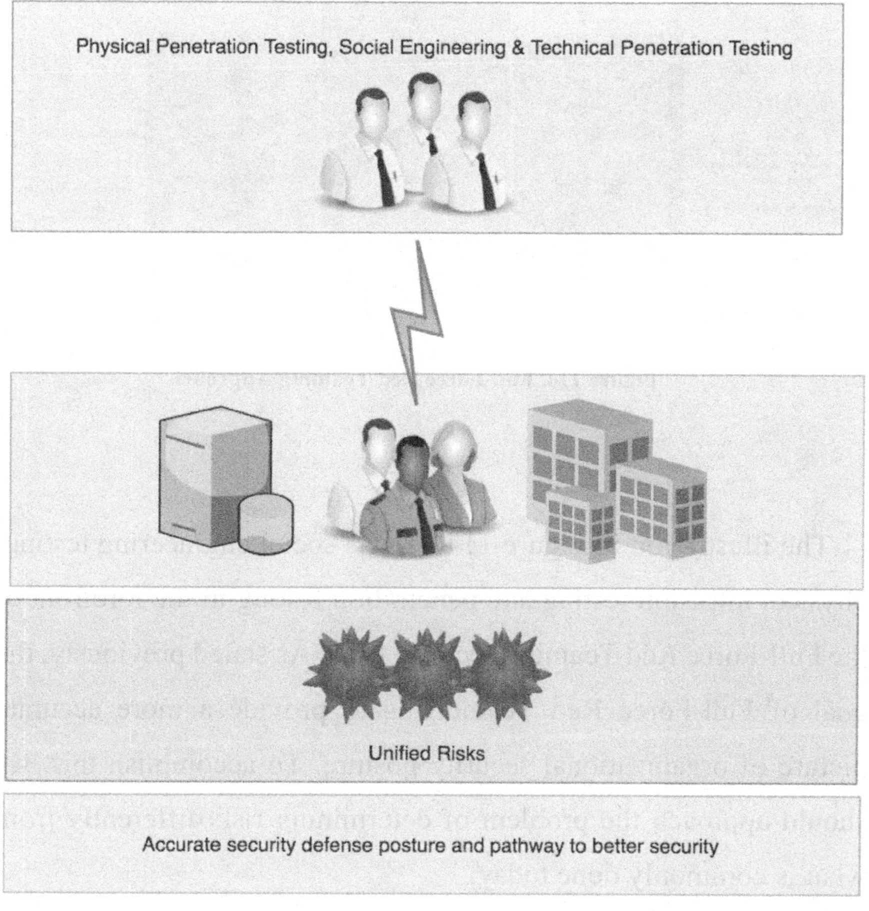

Figure 113. Holistic Offensive Security Testing via Full-Force Red Teaming

As you might have already deduced, a holistic approach toward security testing is a step in the right direction. Full-Force Red Teaming is comprised of comprehensive, correlated security testing on technology, people, and facilities to provide the most accurate security defense posture and a pathway to resilient security.

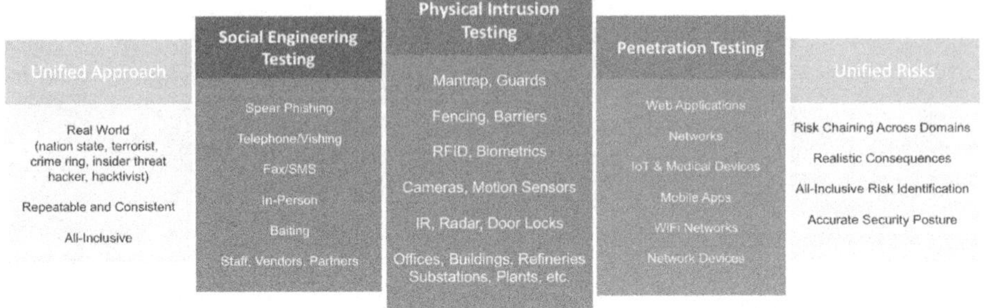

Figure 114. Full-Force Red Teaming Approach

The illustration in Figure 114 depicts social engineering testing, physical intrusion testing and penetration testing in the forefront of the Full-Force Red Teaming security triad. As stated previously, the goal of Full-Force Red Teaming is to provide a more accurate picture of organizational security posture. To accomplish this, we should approach the problem of determining risk differently from what is commonly done today.

So far, we've discussed the absence of correlated physical, social, and technical testing in many of today's organizations. Let us have a closer look at how to approach correlating risk, profiling organizations, and more.

The following section briefly describes one possible approach toward meeting this goal.

One of the initial keys to safeguarding a proper physical red team test is to conduct a profile of the target organization. This profile should take into account the organization from a high level. Aspects that include its industry, size, and so on should be taken into account as part of an overall profile of what threats the organization is likely to encounter. Only then will the operators know which TTPs are relevant and the level of complexity that is needed to ensure a realistic and commensurate test.

That said, profiling an organization is much easier said than done. Here are a few key exposure factors to consider when doing so. Please keep in mind, this is not a complete and comprehensive list, but it will add a layer of depth necessary to root out additional perspective.

Exposure Factors

<u>Industry</u> – Is the organization's industry a likely target? Banking, retail, and hospitality are highly targeted by attackers. Is it in a controversial industry (gambling, abortions, tobacco, firearms)?

<u>Size</u> – How many employees, contractors, partners, agents? More humans generally mean more social engineering target opportunities.

<u>Geographic Footprint</u> – How many offices, stores, places of business? How widespread are physical assets disbursed around the office, building, campus, city, state, country, world? Are they

located in distressed neighborhoods, cities, or countries?

Prominent Characters – Are there any prominent characters associated? Famous people, politicians, high-net-worth families/individuals, or outspoken leaders whose actions/beliefs may introduce additional risk into the environment?

Political Involvement – Does the organization maintain a known political leaning through partisan viewpoints, support, and/or political donations? How does this partisan stance affect the organization?

Customers – Who makes up the majority of customers? Hundreds of business customers? Millions of consumers of everyday goods and services? Do the majority of the customers come from the United States or from unfriendly nations?

Technology Adoption – To what degree does the organization adopt technology into their environment? Does the organization still have Windows XP machines? Is everything in the cloud? How progressive are their technological defenses?

This is by no means a full and complete list. However, the outcome of examining feedback from these exposure factors will ultimately play into building a test plan that addresses likely organizational threats. What's more, it will have a lot to say about the complexity of TTPs to be utilized in order to provide a commensurate and realistic test.

At the very least, running through an exposure factors exercise will provide a clearer picture of the attacker or attackers, what they might be targeting, how they might launch their attack, and their level of sophistication. As a physical red team operator, this information becomes very valuable in crafting an operation of value.

Risk Ranking

It may seem chronologically out-of-step, but let's skip ahead toward the end of the Full-Force Red Team methodology. The process of applying risk to an operation's findings, as you can imagine, is just as critically important as *pwning all the things*. It's definitely not as sexy, but still very important.

Full-Force Red Team methodology calls for testing outside of a vacuum, employing social engineering, physical testing, and technical penetration testing. The next important component is to rank findings not only comprehensively, but as they correlate to one another.

In this section, I will present a sample risk rating methodology that aims to rank findings comprehensively, while correlating with other identified risk aspects in the security triad. Thankfully, this process does not have to differ greatly from what we security pros know as the fundamental equation for evaluating risk.

According to NIST SP 800-30, we have the de facto standard for calculating risk.

RISK = LIKELIHOOD * IMPACT

<u>Impact</u> – The realistic likelihood of threat initiation and successful exploitation

<u>Likelihood</u> – The magnitude of harm to confidentiality, integrity, availability, and accountability of data and resources

<u>Risk</u> – Represents the total amount of risk exposure

There are many ways to evaluate risk exposure by using risk ranking frameworks that adapt to this equation. To keep things simple, I will illustrate the point using a set of granular factors that should be considered when arriving at likelihood and impact. Finally, I'll introduce a quantitative approach my team uses to determine risk.

The image shown in Figure 115 depicts an example of a Likelihood and Impact Table whose purpose is to align with Full-Force Red Teaming. It does this by diving deeper through the use of 12 factors for likelihood and twelve factors for impact. These factors aim to fully represent each side respectively.

Let's take a closer look at Figure 115.

Likelihood		
Bad Actors Objective	0 - 9	
Resources	0 - 9	
Ability	0 - 9	
Immensity	0 - 9	
Organizational Industry	0 - 9	
Size/Employees	0 - 9	
Geographic	0 - 9	
Customers	0 - 9	
Flaws Exploitability	0 - 9	
Detectibility	0 - 9	
Widespread	0 - 9	
Identification	0 - 9	

Impact		
Assets Confidentiality	0 - 9	
Integrity	0 - 9	
Availability	0 - 9	
Traceability	0 - 9	
Customers Monetary	0 - 9	
Reputational	0 - 9	
Privacy	0 - 9	
Litigation	0 - 9	
Organizational Perception	0 - 9	
Monetary	0 - 9	
Compliance	0 - 9	
Employees	0 - 9	

Figure 115. Sample Likelihood and Impact Table

Likelihood and impact are broken down into three categories each (e.g. Bad Actors, Organizational), and each category has a group of four attributes called factors. As you can see from Figure 115, each and every factor has a potential numeric rating of 0 to 9. A rating of 9 indicates the highest estimated measure of presence for any given factor. For instance, a rating of 9 for the factor titled *Ability* in the Bad Actors category for Likelihood would indicate a highly sophisticated level of technical prowess by the attacker. A rating of 0 for the same factor would indicate the attacker is likely to have virtually no technical aptitude, and so on.

Likelihood & Impact Factors

Before we go any further, let's extrapolate the factors in Figure 115 to understand how they work. Each of the categories and their respective factors have been developed with the sole purpose of capturing risk on a grander level. To do that, twenty-four factors are broken up into six categories help make that happen.

To start, let's unpack each of the factors by first beginning with Likelihood and then provide guidance on how to use them.

Likelihood		
	Factor	**Description**
Bad Actors	Objective	How substantial is the reward? How motivated is the attacker?
	Resources	How well-resourced is the likely attacker?
	Ability	How skilled is the likely attacker?
	Immensity	How large is the group of attackers?

Table 1. Likelihood Risk Factors Explained for Bad Actors

Likelihood		
	Factor	**Description**
Organizational	Industry	How common is this threat to this industry?
	Size/Employees	How large is the organization?
	Geographic	How geographically dispersed is the footprint?
	Customers	How big is the base of clients, customers, partners?

Table 2. Likelihood Risk Factors Explained for Organizational

Likelihood		
	Factor	**Description**
Flaws	Exploitability	How easy is the exploit to carry out?
	Detectability	How likely is a successful exploit to be detected?
	Widespread	How widely known is this vulnerability?
	Identification	How easy is this threat to be identified by attackers?

Table 3. Likelihood Risk Factors Explained for Flaws

As shown in Figure 115, each and every factor has a potential numeric rating of 0 to 9. A rating of 9 indicates the highest estimated measure of presence for any given factor. We'll cover more on calculating risk later in this section. For now, let's unpack each of the risk factors for Impact.

See Tables 4 through 6.

Impact		
	Factor	**Description**
Assets	Confidentiality	How much could be lost (data, assets) and its sensitivity?
	Integrity	How much corrupt or "dirty" data?
	Availability	How widespread and vital could be the loss of service?
	Traceability	How traceable to an individual would the actions be?

Table 4. Impact Risk Factors Explained for Assets

Impact		
	Factor	**Description**
Customers	Monetary	What would the monetary impact be to customers?
	Reputational	What would be the reputational damage done to customers (financial, personal)?
	Privacy	What would be the impact on personal information (PII, healthcare)?
	Litigation	What would be the impact for litigation from customers?

Table 5. Impact Risk Factors Explained for Customers

Impact		
	Factor	**Description**
Organizational	Perception	How much harm to the org's brand may happen?
	Monetary	How much financial damage will the org incur?
	Compliance	How will this impact compliance (HIPAA, PCI, FISMA)?
	Employee	How much damage to the org's employees?

Table 6. Impact Risk Factors Explained for Organizational

The information in Tables 1 through 6 are intended to help weigh risks with greater depth. But like any system, they are not perfect.

While I believe this is a dramatic improvement over what's being practiced by red teams today, I advise your team to adopt and revise as necessary.

Now that we are familiar with the Likelihood and Impact table (Figure 115) and know what the risk factors mean (Tables 1 to 6), we are ready to dive into the next leg of our journey, risk calculation.

Risk calculation is about as straightforward as it gets – there's no need to get fancy here. You may have a need for more complexity, but most of us do not. In fact, I've found that many red teams do not have a risk calculation process at all. They skip the math and rely on the purely subjective approach. If you're one of the subjective purists, you probably started cringing at the sight of Figure 115. Sorry, I'm not sorry. Call it a personal preference, but a subjective approach won't do for me, and from experience, it often won't do for my clients either. I am much more productive by spending a lot less time debating with clients the infinitesimal nuances of why a finding should be considered a medium instead of a high. That said, a quantitative process is not absent of at least *some* subjectivity. However, a formulaic process does offer a pathway to repeatability and consistency, if even from a perception perspective.

Moving on to risk calculation, let's take Figure 116 into consideration. The calculation total for both Likelihood and Impact are comprised of an equation of averages. Each of the values for the twelve factors for Likelihood would be added up and divided by 12 to arrive at the total average. The same is done for the twelve factors for Impact.

Calculating the Likelihood and Impact for a finding looks like this:

Likelihood		
Bad Actors	Objective	9
	Resources	4
	Ability	9
	Immensity	9
Organizational	Industry	4
	Size/Employees	7
	Geographic	5
	Customers	5
Flaws	Exploitability	8
	Detectibility	4
	Widespread	4
	Identification	4
	Sum All Factors	72

Total Likelihood 6.00
For example: 57 / 12

Impact		
Assets	Confidentiality	4
	Integrity	4
	Availability	9
	Traceability	2
Customers	Monetary	1
	Reputational	9
	Privacy	5
	Litigation	3
Organizational	Perception	8
	Monetary	8
	Compliance	5
	Employees	2
	Sum All Factors	60

Total Impact 5.00
For example: 57 / 12

Figure 116. Risk Analysis Calculation Example

As you can see in Figure 116, some sample numbers have been thrown in for demonstrative purposes. The sum of each of the twelve factors have been averaged together where Total Likelihood is 6.00 and Total Impact is 5.00. Now that we have the numeric values, let's add a little more context by applying their risk values to another table to determine their risk labels.

Likelihood and Impact Levels	
0 to <3	LOW
3 to <6	MEDIUM
6 to 9	HIGH

Figure 117. Likelihood and Impact Labels

Taking Figure 117 into account, we can see that the risk label for Likelihood in our example in Figure 116 would be **MEDIUM,** because its risk value is 6.00. Similarly, the risk label for Impact in our example would also be **MEDIUM,** because its risk value is 5.00.

Now that we have the risk value and labels, we can calculate the overall risk value and label in a similar manner. We arrive at our Overall Risk Value by adding the risk values for Likelihood and Impact, in our sample finding, and then divide by two. As a result, we find ourselves with an Overall Risk Value of 5.50.

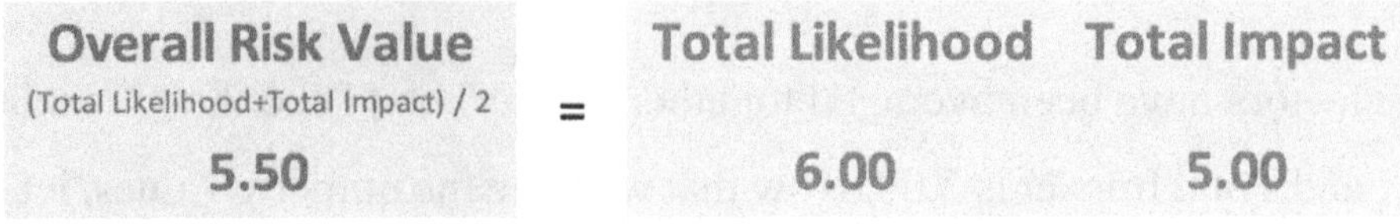

Figure 118. Overall Risk Value Calculation

Please see Figure 118 for the formula to calculate Overall Risk Value.

With the finding's Overall Risk Value in hand, 5.50, we can now find its Overall Risk Label by referring to the table in Figure 119.

Overall Risk = Likelihood x Impact				
	HIGH	Medium	High	Critical
Impact	MEDIUM	Low	Medium	High
	LOW	Note	Low	Medium
		LOW	MEDIUM	HIGH
	Likelihood			

Figure 119. Overall Risk Table

It is easy to see that in this example, the Overall Risk Label is **MEDIUM**, since both Likelihood and Impact are **MEDIUM**. If the labels were different, you would merely find the risk label in Figure 119 for Impact vertically and match it to the corresponding Likelihood risk label horizontally. Simple, yes? This risk ranking approach is not perfect for everyone, but it offers a better solution to most these days. In fact, one could achieve greater granularity by weighting the likelihood and impact factors to their liking. I suggest starting simple and adding complexity only when necessary.

To summarize, each finding discovered as part of the physical red team operation should undergo this risk analysis exercise to determine a clearer risk value and label. I recommend using this methodology or a modified version of it to substantiate findings and ensure consistency across the board.

For hands-on Physical Red Team Training, please visit:

https://www.redteamsecuritytraining.com/physical-red-teaming-book

[EXTRA]

ABOUT THE AUTHOR

Jeremiah has been in the information technology industry for over 20 years and is the creator of The PlugBot Research Project – a foray into the concept of a hardware botnet. He has a master's degree in Information Security & Assurance and an executive business education from the University of Notre Dame, Mendoza School of Business.

Jeremiah is the founder and chairman of RedTeam Security, a St. Paul, Minnesota-based information security consulting firm specializing in physical red teaming, penetration testing, and social engineering. He is the founder and principal instructor for Red Team Security Training, an advanced training facility focusing on physical red teaming and other forms of offensive security training.

Jeremiah is the author of *The Social Engineer's Playbook: A Practical Guide to Pretexting* and has appeared on the national TV news outlet, FOX News Live with Adam Housley. Additionally, he has been quoted and interviewed by CNN, Yahoo! Finance, Houston Chronicle, Rigzone, Business Insider, Tech Insider, CIO Magazine, PenTest Magazine, POWER Magazine, and many others.

Jeremiah has served as CISO and expert consultant to several Fortune 500 companies. He is a CISSP, CCISO, CEH, CHFI, and CCENT. He is the author of two patents and a former adjunct professor at Norwich University, College of Graduate Studies in Information Security & Assurance.

Additional Resources:

Personal
Website – http://www.jeremiahtalamantes.com
LinkedIn – http://www.linkedin.com/in/jtalamantes

Red Team Security Training
Website – https://www.redteamsecuritytraining.com
Facebook – https://www.facebook.com/redteamtraining
YouTube – https://bit.ly/2vbQqD9
Twitter – https://twitter.com/redteamtraining

Red Team Security
Website – https://www.redteamsecure.com
Facebook – https://www.facebook.com/redteamsecure

Made in the USA
Monee, IL
07 July 2026